Grafting & Budding

Grafting & Budding

SECOND EDITION

A Practical Guide for Fruit and Nut Plants and Ornamentals

W J Lewis and D McE Alexander

LAND LINKS

© Bill Lewis 2008

All rights reserved. Except under the conditions described in the *Australian Copyright Act* 1968 and subsequent amendments, no part of this publication may be reproduced, stored in a retrieval system or transmitted in any form or by any means, electronic, mechanical, photocopying, recording, duplicating or otherwise, without the prior permission of the copyright owner. Contact CSIRO Publishing for all permission requests.

National Library of Australia Cataloguing-in-Publication entry
   Lewis, William J.
   Grafting and budding: a practical guide for fruit and nut
   trees and ornamentals/W. J. Lewis, D. McE. Alexander.

   2nd ed.

   9780643093973 (pbk.)

   Includes index.
   Bibliography.

   Fruit trees.
   Nut trees.
   Plants, Ornamental.
   Grafting.
   Budding (Plant propagation)

   Alexander, D. McE. (Donald McEwan)

   634.0441

Published by
Landlinks Press
36 Gardiner Road, Clayton VIC 3168
Private Bag 10, Clayton South VIC 3169
Australia

Telephone: +61 3 9545 8400
Email: publishing.sales@csiro.au
Web site: www.publish.csiro.au
Sign up to our email alerts:
publish.csiro.au/earlyalert

Landlinks Press is an imprint of **CSIRO** Publishing

Front cover
T-budding, first cut on the rootstock. Photo by Bill Lewis.

Back cover (from left)
Grafting tools; protecting an avocado scion from drying; making a grafting cut.
Photos by G Brown.

Set in 12/15.5 Adobe Minion
Edited by Janet Walker
Cover and text design by James Kelly
Typeset by Desktop Concepts Pty Ltd, Melbourne
Printed by Ingram Lightning Source

CSIRO Publishing publishes and distributes scientific, technical and health science books, magazines and journals from Australia to a worldwide audience and conducts these activities autonomously from the research activities of the Commonwealth Scientific and Industrial Research Organisation (CSIRO). The views expressed in this publication are those of the author(s) and do not necessarily represent those of, and should not be attributed to, the publisher or CSIRO. The copyright owner shall not be liable for technical or other errors or omissions contained herein. The reader/user accepts all risks and responsibility for losses, damages, costs and other consequences resulting directly or indirectly from using this information.

CSIRO acknowledges the Traditional Owners of the lands that we live and work on across Australia and pays its respect to Elders past and present. CSIRO recognises that Aboriginal and Torres Strait Islander peoples have made and will continue to make extraordinary contributions to all aspects of Australian life including culture, economy and science. CSIRO is committed to reconciliation and demonstrating respect for Indigenous knowledge and science. The use of Western science in this publication should not be interpreted as diminishing the knowledge of plants, animals and environment from Indigenous ecological knowledge systems.

Jun25_RP_ILS

Acknowledgments

The authors thank the late Mr E.A. Lawton, Mr G. Brown, Bob Merlin, Shona Duffield and Narie Lewis for the photographs and Mrs K.I.B. O'Grady and Associate Professor L. Halvorsen for editorial assistance. Thanks are also due to colleagues at CSIRO Plant Industry, Horticulture Unit.

The following nurseries must be thanked for their help with information for the new sections on individual plants:

- Australian native plants – Humphris Nursery, Lang's Native Plant Nursery, Bushgrafts Nursery
- Cacti – Paradisia Nursery
- Conifers – Coolwyn Nurseries
- General deciduous ornamentals – Southern Weepers and Ornamental, Fleming's Nurseries
- Passionfruit – Nellie Kellie Passionfruit, Birdwood Nursery
- Roses – Monbulk Rose Farm, Corporate Roses, Roworth's Rose Nursery
- Tomatoes – Oasis Nursery, Plummer's Nursery, Flavorite Tomatoes

Contents

	Acknowledgments	v
1	**Introduction**	1
2	**Grafting – the basics**	7
	The grafting cut	7
	Scion wood	9
3	**Budding techniques**	13
	T-budding	13
	Chip budding	19
	Patch budding	20
	V-budding	23
4	**Grafting techniques**	25
	Splice or whip graft	26
	Wedge or cleft graft	28
	Whip and tongue graft	30
	Bark graft	32
	Side graft	33
	Approach graft	35
5	**Methods for selected species**	39
	Australian native plants	39
	Annonas or custard apples	47
	Avocado	49
	Cacti	52
	Cashew	56
	Citrus	59
	Conifers	61
	General deciduous ornamentals	65
	Grape	68

Macadamia	70
Mango	71
Passionfruit	73
Pistachio	77
Pome fruit (apples and pears)	79
Roses	80
Sapodilla	82
Stone fruits	85
Tomatoes	86
Walnut	91
Ziziphus	92
Glossary	*95*
Further reading	*97*
Internet resources	*99*
Index	*101*

Introduction

From the beginnings of horticulture, growers have tried to improve their orchards and gardens by choosing and keeping the productive or desirable trees. At first, there were merely stands of seedling trees, either natural or planted. Each tree was different from the others, as usually happens when plants are grown from sexually produced seeds.

Eventually growers discovered how to make almost exact copies of their superior plants. One way is by taking cuttings. Put simply, you cut part of a shoot from the original plant and stick it into the ground where it grows some roots.

Another way is by grafting. Once again, you cut part of a shoot from the original superior plant. In grafting, this part is called the scion. You then attach this scion to another plant of the same sort (the rootstock) in such a way that the two unite and grow together. Shoot growth from the rootstock is discouraged, so that the scion grows to become the trunk and branches. The rootstock, as the name suggests, provides only the roots and perhaps a short part of the lower trunk. These asexual methods of reproduction are called vegetative propagation. The cuttings, and the scion portions of plants made by grafting, are clones of the parent plant.

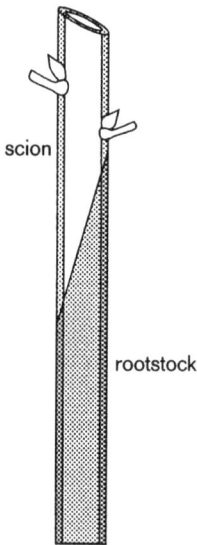

Figure 1. A graft consists of a rootstock and a scion.

These days, commercial fruit and nut trees are nearly always propagated vegetatively from selected varieties, as are many important ornamental plants.

Plants produced in this way maintain all of the characteristics for which the variety was selected, such as better fruit yield and quality, desirable form, flowers and foliage, and resistance to pests and diseases. By contrast, seedlings of most plants are inherently variable. Even if the parent plant is a selected variety, its seedlings will vary in crop production and quality from one to another and will rarely equal the parent plant in all aspects. A further disadvantage of seedling fruit trees is that they often take much longer to first crop.

To make a graft, the rootstock must unite successfully with the desired scion. RJ Garner, in his *Grafter's Handbook* (see Further reading), described the formation of the graft union as '*the healing in common of wounds*'. This healing in plants begins with the formation of scar tissue or callus, produced either from the plant's cambium, or from nearby immature wood and bark cells. The cambium is the thin (cylindrical) layer of cells between the bark and wood, where annual growth originates

Figure 2. A seedling sweet orange rootstock plant suitable for budding.

in woody plants. Whenever a scion or rootstock shoot is cut, the cambial layer is exposed along the cut surfaces where the wood meets the bark. All grafting techniques must achieve intimate contact between the cambial regions of the scion and stock, so that they can grow together.

Compatibility between rootstock and scion exists only between plants of the same species or closely related plants. Seedlings of the same species as the scion are often used as rootstocks. Seedlings are cheap and easy to produce and sometimes have better root anchorage than cuttings.

However, seedlings are variable, so increasingly, rootstocks are also propagated vegetatively as cuttings. For many major crops, rootstocks have been selected for resistance or tolerance to soil-borne problems such as fungi and nematodes, different soil types, drought, and waterlogging. Once compatibility has been considered, the rootstock can be chosen, independently from the scion, to suit the planting site.

Most modern fruit varieties (and some ornamental plants) are propagated by grafting selected scions onto selected rootstocks. However, a few

Figure 3. A crepe myrtle accidentally ringbarked by a line trimmer was saved by bridge grafting. Shoots arising below the ringbark were grafted under the bark above the ringbark.

species such as grapevines, figs, olives and some roses grow successfully as cuttings and are propagated commercially in this way.

This book describes techniques of budding (bud grafting) – where the scion is a single vegetative bud with only a small piece of stem attached – as well as grafting, where the scion is a length of shoot bearing one or several vegetative buds.

The goal of the grafting process is to form a union between the tissues of the scion and those of the rootstock, followed by suppression of any shoot growth from the rootstock, so that the scion becomes the new aerial part of the plant. The success of the operation depends on having a rootstock in prime condition, selecting quality scion wood and using the most suitable technique for budding or grafting. The timing of the operation and correct after care are also important ingredients for success.

In the 1970s, the CSIRO Division of Horticultural Research (now CSIRO Plant Industry, Horticulture Unit) established a wide range of tree fruit species at test sites throughout Australia. This book describes the techniques used for propagation of these various deciduous and evergreen species. It provides examples of the selection and storage of scion wood, the timing and techniques of the budding or grafting operation, and suggests a timetable for each step of the procedure.

The techniques described in this book are widely used in the commercial production of grafted plants. Home garden enthusiasts can also use these methods to multi-graft their fruit trees, thus making best use of their available space.

A multi-grafted tree consists of several scion varieties worked onto a single rootstock. Generally, different plants that are closely related may be grown on a single rootstock, such as most citrus varieties, most stone fruits, or a selection of pome fruits. Similarly, different cultivars of a single species can be grown on one tree by budding different varieties to different branches. For example, a range of peach varieties grafted on one tree will mature their fruit from November to March in southern Australia.

However, scions for multi-grafting should be carefully selected to ensure success. If virus-free scion wood is not available, virus or virus-like

Figure 4. A multi-grafted male pistachio tree.

Figure 5. The essential tools for grafting.

diseases may cause stunting and interfere with the balanced growth of the tree.

In addition, you should choose scion varieties that have similar vegetative vigour, or else the strongest variety will soon overgrow the others.

Grafting – the basics

The essential tools you will need for budding or grafting are:

- secateurs for making rough preliminary cuts;
- a very sharp knife for making the grafting cuts;
- some suitable binding material, such as plastic budding tape;
- some wound sealant, or scion covers, such as small plastic bags.

Professional grafters have special budding or grafting knives, but you can use any knife of a convenient size and shape. Knives with disposable or replaceable blades avoid the need for sharpening. Be careful when using sharp tools, and check that your fingers are out of harm's way before making cuts.

The grafting cut

Well-made grafting cuts result in flat, smooth cut surfaces of stock and scion that will lie together with maximum contact, especially where the cambial layer is exposed. For many grafting and budding cuts, you will need to move the knife towards your body, an action that contradicts the usual safety advice for handling sharp knives. You should draw the knife

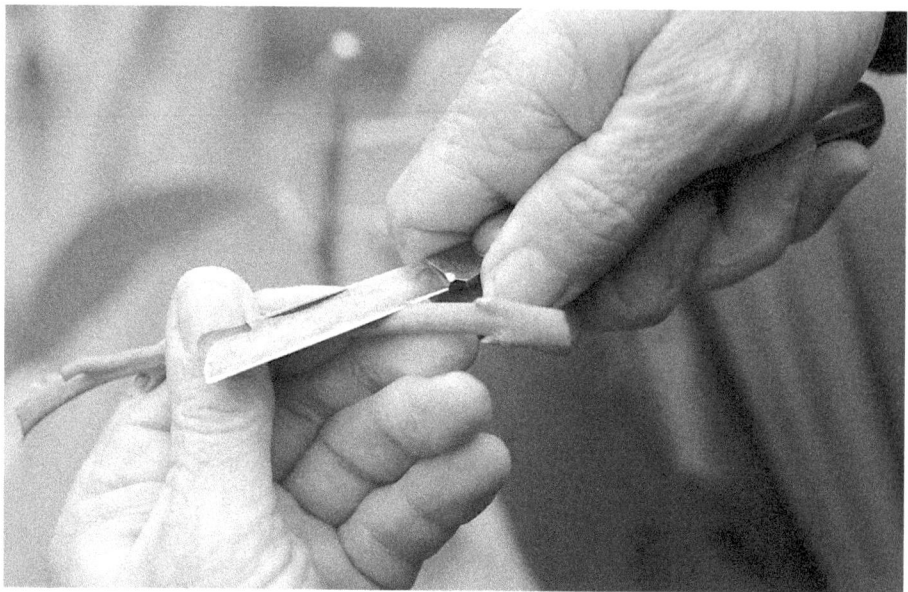

Figure 6. Making a grafting cut.

through the wood with a slicing action, rather than trying to push it straight through (see Figure 6).

If you are an inexperienced grafter, you should practise your cuts on some prunings of the intended scion or rootstock.

1. Take the knife in your dominant hand, with the sharp edge facing your body.
2. Hold the scion (or support the rootstock) with the other hand, on the far side of the point of attack.
3. Lay the knife blade flat on the nearest surface of the shoot, diagonally across it.
4. Position the blade on the shoot so that the handle end of the blade is at the desired starting point.
5. The knife should now be in such a position that the butt of the handle is closest to you, and the tip of the blade is facing diagonally away. The angle between the sharp edge of your knife and the shoot will be 30–60 degrees, with the blade still flat on the surface of the shoot.

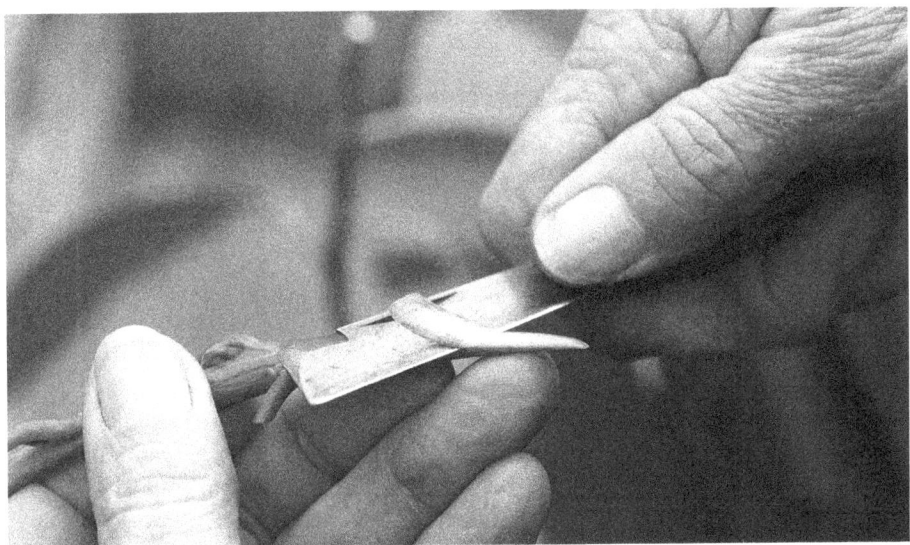

Figure 7. Paring off a thin slice to improve a grafting cut.

6. Raise the back edge of the blade slightly from the surface of the shoot to allow the cutting edge to bite into the bark.

7. Make the cut by drawing the blade along, through the shoot, and at the same time across the shoot, so that the length of the blade is used, moving from base to tip as you cut through the shoot.

8. Make the whole of the cut in one smooth stroke. Stopping or sawing will inevitably result in a rough or jagged cut.

Often it is easier to get a good flat cut on the second try, when you are only paring off a thin slice rather than cutting through the full stick. Therefore, you should start by cutting the scion or stock just a little longer than you want it.

Scion wood

Take scion material from sources that are true to variety, that have superior performance, and if possible, from plants of known good health. Always select solid shoots that are well exposed to sunshine and avoid

Figure 8. An avocado scion protected from drying with a plastic bag.

weak, shaded shoots. If you cut scion wood from a grafted source tree, watch out for rootstock suckers.

It is best to collect and use scion wood on the same day, but you can store wood from plants native to cool and temperate areas in a sealed plastic bag in an ordinary refrigerator. Maintain humidity inside the bag by using dampened newspaper as a wrapping. Under these conditions, scion wood will keep for at least several weeks.

If you wish to store propagating material of plants from tropical zones, keep it in a cool room at about 10°C. Storage in a refrigerator (around 5°C) will cause cold damage.

When collecting scion wood, cut wood intended for budding into budsticks of convenient lengths (100–150 mm), each bearing 5–10 buds. Cut wood for grafting as individual scion pieces of 50–100 mm with one or more buds, or as multiples of this length. The best type of wood varies with its intended use, and this aspect is discussed in later sections on different techniques for budding and grafting.

Rootstocks ready to be budded or grafted should be healthy and well watered if in leaf or evergreen. They should have reached such a size that the stem is about pencil thickness (5 mm diameter) at the intended position of working, 10–15 cm above the roots.

Rootstocks are usually trained to a single main shoot by regularly pinching out any lateral shoots. For the purpose of description in this book, it has been assumed that rootstock shoots are growing vertically although this may not always be so.

With all budding and grafting techniques, you should prepare the rootstock first and then the scion, so that a minimum of time elapses between cutting the scion and its final wrapping. Plan the operation beforehand, so that tools, scion wood, and tying materials are laid out conveniently. Any delays may allow the cut surfaces of scion and stock to dry out, reducing the chance of success. In dry climates, perform the whole grafting operation in a humid location that is protected from the wind.

For a successful union, the care of plants after budding or grafting is as essential as the operation itself. The following four points are of prime importance:

- Remove or suppress all shoot growth from the rootstock as soon as it appears, so that the scion is not overgrown by the rootstock.
- Temperatures should be between 15° and 30°C for good growth and callus formation. (Temperatures above 30°C or below 10°C can slow down or prevent callus growth so that scions die or take an unnecessarily long time to develop.)
- Supply adequate (but not excessive) water and nutrients to maintain healthy vegetative growth.
- Remove or loosen the budding or grafting tape before any growth restriction occurs.

Budding techniques

Budding is a special type of grafting, in which a small piece of shoot carrying a single vegetative bud is sliced from the scion wood and transferred to the rootstock. Usually, buds used for budding are found in the axils of leaves, between the leaf stalk (petiole) and the shoot, on the side of the petiole away from the base of the shoot (see Figure 16). This fact is useful in determining the orientation of the bud after it has been cut, as it will grow much better if it is placed on the rootstock the right way up! Even on dormant wood from deciduous plants, you can see the scar where there was once a leaf (see Figure 14).

Budding has two advantages over grafting. First, less scion wood is used, since only one bud is needed per plant, and second, the budding operation is quicker than grafting. Common budding techniques include T-budding, chip budding, and patch budding. With each technique, several variations are available to suit individual operators and different conditions.

T-budding

T-budding gets its name from the shape of the cuts made in the rootstock bark to prepare it for the insertion of the scion bud. It is also called shield

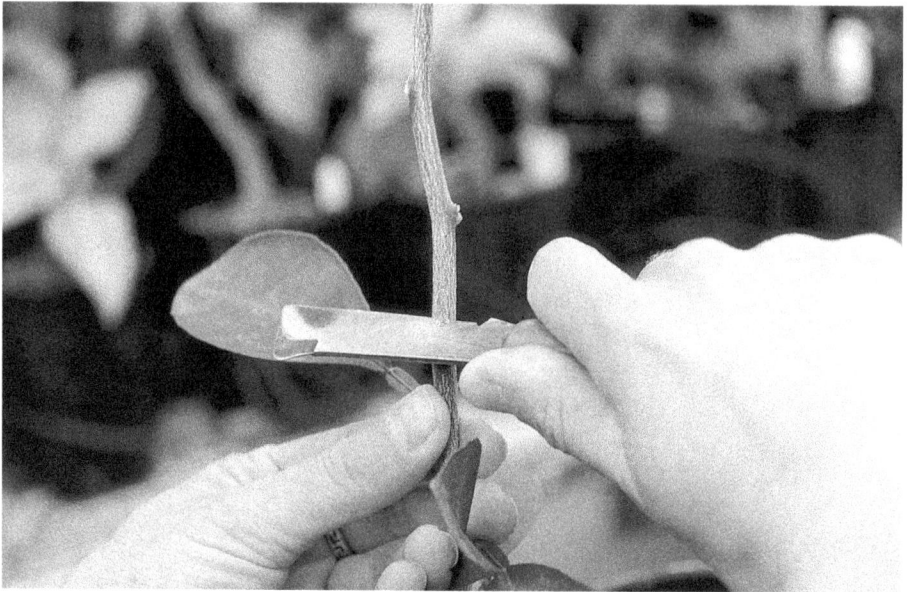

Figure 9. First cut across the rootstock bark for an upright T.

budding, because of the shape of the small piece of scion shoot transferred with the bud. T-budding is the quickest method of budding or grafting, so is widely used in the production of large numbers of budded citrus and roses.

You can T-bud only when the rootstock is in active growth and the bark separates easily from the wood. Check with a thumbnail, or the point of a knife. If the bark does not lift or slip easily from the wood, try chip budding, or one of the grafting techniques, or wait until the bark is slipping.

The most suitable scion wood is the middle third of vigorous shoots of the last mature flush of growth. This wood has partly hardened and has filled out, so that it is more rounded and less angular in cross section. As you collect it, cut off all the leaves, leaving a 5 mm stub of each petiole. If the bud wood is to be stored for more than a few weeks, you should seal the cut ends of the shoots with a low melting point wax or a water-based plastic paint.

To prepare the rootstock, first select a straight part of the stem at the desired position. Trim off any leaves or thorns that would interfere with your work.

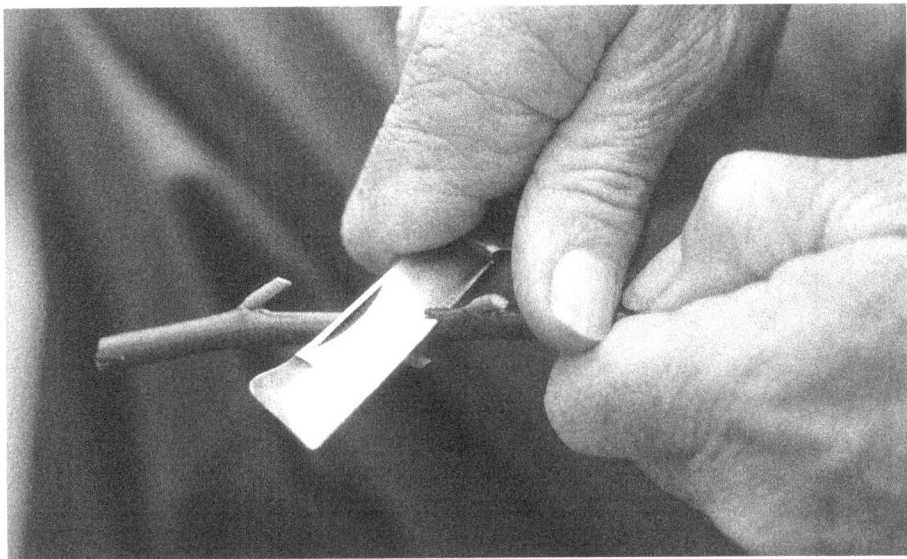

Figure 10. Cutting a bud for an upright T.

When you use your spare hand to steady the rootstock shoot while making the T cuts, avoid possible injury by holding the stem above the area of operation and not behind it.

To perform an upright T-bud: make the first cut across the rootstock shoot, through the bark, but not into the wood. This cut should extend about one-third of the way around the shoot and can be either straight or slightly curved with the convex side upwards. While making the cut, hold the knife so the flat of the blade is at about a 45 degree angle above the horizontal. The curved, angled cut makes it easier to lift the bark. The second cut, also just through the bark, runs from the centre of the first cut straight down the shoot for about 3 cm. Starting from the junction of the two cuts, lift a flap of bark from the wood along both sides of the second cut. Ease the bark away from the wood using either the tip of the knife or the special attachment provided on a budding knife. After lifting, in order to minimise drying of the exposed tissues, you can press the bark temporarily back into place while you prepare the bud.

Select a bud stick and hold it by its upper end, with the basal end facing away from your chest. Rest the thumb of your knife hand on the upper surface of the bud stick, just above the bud to be cut. The sharp edge of the knife blade is laid across the scion stick at an angle, as described for grafting. Grasp the knife handle in your partly clenched fingers. Make a shallow cut starting about 1 cm below the bud, passing under it, then exiting about 1 cm above it. In making the cut, move the knife by further clenching your hand towards the thumb, in a fashion similar to the action of peeling potatoes, and drawing slightly rather than pushing straight through. Make the cut under the bud in a single stroke, so that the cut surface of the bud piece is reasonably smooth.

An alternative way to take the bud is simply to stop the shallow cut at about 1 cm above the bud, then separate the bud piece by a second small cut across the budstick at that point.

Some operators remove the small sliver of wood from the underside of the bud piece, but that is not essential.

Quickly insert the bud piece beneath the lifted rootstock bark at the top of the T, with the petiole side of the bud towards the base of the rootstock. Ease the bud piece gently down into position until it is all under the bark, with only the bud and petiole stub protruding (Figure 13). Use the petiole

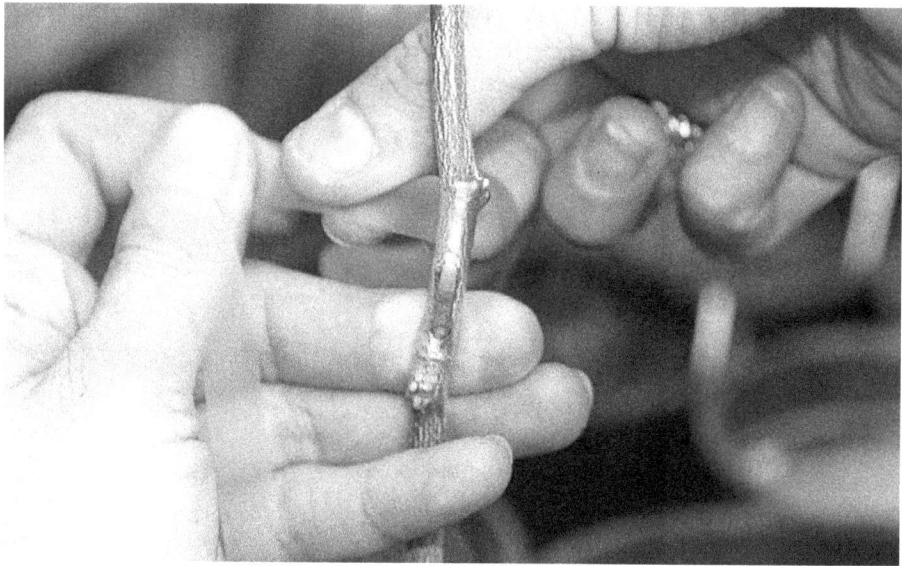

Figure 11. Passing the tape around the stem.

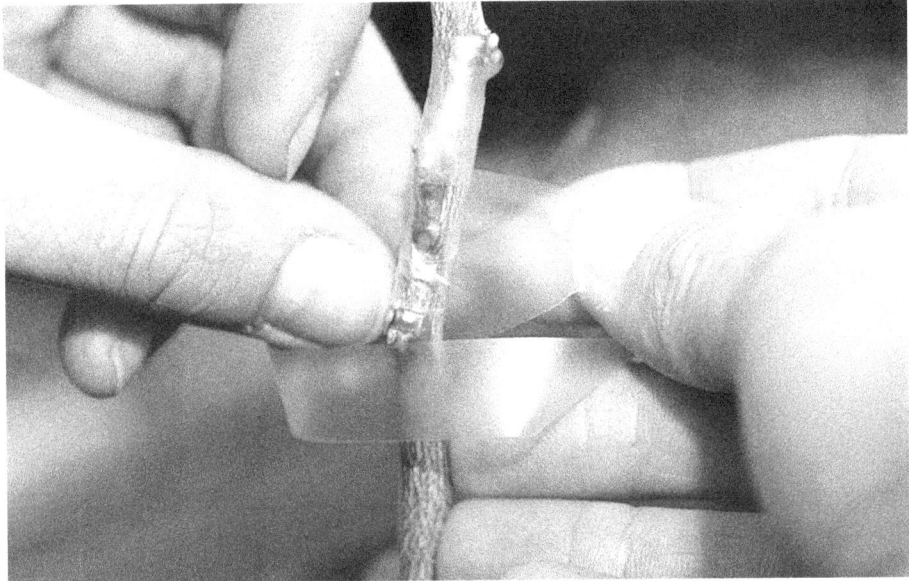

Figure 12. Passing the end of the tape through the last turn to tie off.

stub as a handle for this operation. If the bud piece is a little too long, cut off the exposed tail at the cross cut of the T.

Complete the operation by binding the bud piece firmly into place. Although some budders use rubber strips or patches, the most readily available binding material is 12 mm wide PVC budding tape. Break off a 20–25 cm length of budding tape. Start wrapping just above the top of the T by grasping one end of the tape between the thumb or forefinger of one hand and the stock. With the other hand, wrap the tape tightly around the stock and over the held end of the tape to secure it. Continue wrapping in an overlapping, spiral fashion, down past the base of the T and back up again.

Keep as much tension in the budding tape as you can, to the point that it stretches slightly as you wrap it. Hold the stock plant with one hand and tension the tape with the other. You need to pass the tape from hand to hand to go around the rootstock shoot, but you also need to hold the rootstock shoot so that you can pull on the tape as you are wrapping it. So, if you are right handed, hold the rootstock shoot with the last three fingers of your left hand. This leaves your thumb and forefinger free for passing the tape from hand to hand around the shoot. Wrap the tape mainly with your right hand, with your left thumb and forefinger just holding the tape and maintaining the tension while you move your free right hand back around the shoot to pick up the tape again for another turn.

Cover the bud and petiole stub completely, or leave the bud exposed between the turns of tape. Reduce the tape tension if you pass over the eye of the bud. Secure the end of the tape by slipping it under the last turn and pulling it tight.

The first sign that the bud has successfully taken is when the petiole stub turns from green to yellow. Soon after that, the petiole stub separates from the bud shield, and then you will notice the bud starting to swell and push out. If you covered the bud, partially unwrap the tape after two to three weeks so that bud can grow without restriction.

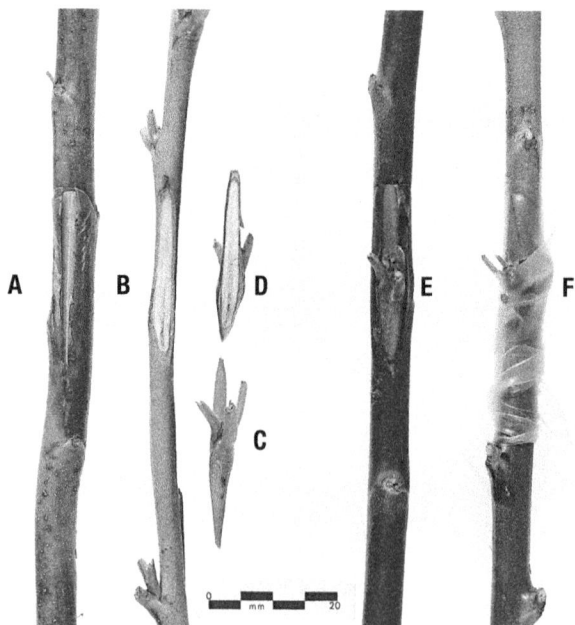

Figure 13. T-bud of peach. A) Bark of rootstock prepared. B) Bud cut from selected scion. C) Outer view of bud. D) Inner view of bud. E) Bud inserted in rootstock. F) Bud wrapped with PVC budding tape.

At the time of budding, prune off the upper third of the rootstock shoot.

After the new bud sprouts, further shorten the top of the rootstock, or cut half way through it 1 cm above the bud and bend it over (see Figure 37D). In windy situations, leave a 10 cm stub of rootstock above the bud as a support, to tie up the new shoot during its early growth. Completely prune off the stub in the following year. Pinch out any bursting rootstock buds as they appear.

Completely remove the tape after about three months, or earlier if necessary, to prevent any restriction to growth. A rapid and convenient method of removing the tape is to slit it down one side with a sharp knife.

With some plants, such as citrus, the inverted T-bud is more commonly used.

As the name suggests, the incisions in the rootstock bark are performed in the shape of an upside-down T, the scion bud is cut from the budstick starting from above the bud, and the bud is inserted into the T from below.

Another modification of T-budding, known as microbudding, has also been used for citrus. Younger rootstocks are used and bud wood is selected from less mature shoots, which are still angular in cross section rather than round. The bud piece and T-cuts are made smaller than for normal T-budding. Microbudding requires more dexterity, but has the advantage that the less mature scion and rootstock materials grow together more quickly, and a budded tree can be produced earlier. The younger bud wood is often more readily available, but it has a limited storage life and is best used fresh.

Chip budding

Chip budding is used when it is not possible to T-bud because the rootstock bark is not lifting, perhaps due to unsuitable growing conditions or seasonal dormancy. Chip budding may also be more successful than

T-budding in cooler climates, where callusing is slower. As with T-budding, choose scion wood that is about the same diameter as the rootstock, or slightly smaller.

This technique is also quick because it involves only two knife cuts on both the rootstock and scion.

Prepare the rootstock as follows:

1. Select a smooth, straight piece of stem with no buds.
2. Make the first cut directly across the stem, angled down at about 30 degrees towards the base of the stem and passing about one-quarter of the way through it.
3. For the second cut, start about 2 cm above the first cut, angle in at first, then down to meet the bottom of the first cut. Discard the chip (Figure 14A).
4. Make two similar cuts in the scion, but with a bud in the middle of the chip (Figure 14B, C and D).
5. Quickly fit the scion chip into the rootstock, so that the cambial areas of the scion and stock match on at least one side, or preferably both.
6. Finally, bind the chip tightly into place with budding tape as described for T-budding. (Figure 11).

With species such as pistachio, growing buds at the ends of the shoots tend to suppress lower buds. In this case, you can improve scion bud burst by semi-cincturing the rootstock about 1 cm above the inserted bud. Make two parallel knife cuts just through the bark, 3–5 mm apart and passing halfway around the stem. Remove the small strip of bark between the cuts.

Patch budding

This technique is used for species that have thick bark, or bark that tends to split along the stem, such as walnut and cashew, and for species that produce latex, such as sapodilla and jackfruit.

Budding techniques | 21

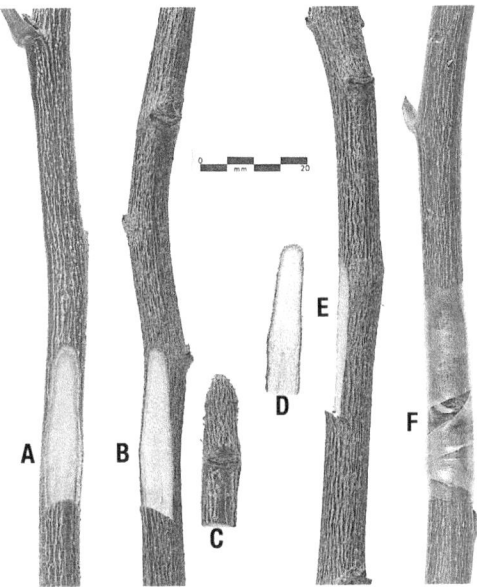

Figure 14. Chip bud of pistachio. A) Rootstock prepared for chip bud. B) Bud cut from selected scion. C) Chip bud outer view. D) Chip bud inner view. E) Bud cut from selected scion. F) Chip bud inserted and wrapped.

The technique consists of removing a rectangular patch of bark from a rootstock and replacing it with a similar sized patch of scion bark carrying a single bud, as illustrated in Figure 15. The bark must be slipping (that is, easily removed from the wood) on both stock and scion. To induce bark slip in scion wood that has been stored under refrigeration, hold it at about 20°C for two to three weeks prior to budding. During this time, stand the sticks with their bases in water.

Choose a straight portion of rootstock stem with no buds. Make two parallel cuts around the rootstock stem, about 3 cm apart and 1–2 cm long, just through the bark. Next, make two parallel cuts along the stem, joining the ends of the first two cuts, so that a rectangle of bark can be removed.

Cut a similar sized patch around a plump vegetative bud on a scion budstick of similar diameter to the rootstock. It is important that the scion patch fits exactly between the top and bottom cuts on the rootstock, but the fit along the sides is not critical. Slip the scion patch from the

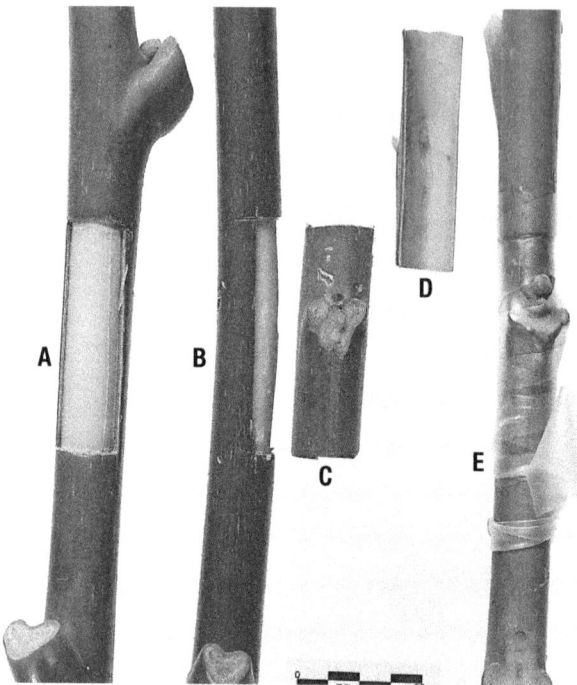

Figure 15. Patch bud of walnut. A) Rootstock prepared for patch bud. B) Patch bud cut from selected scion. C) Patch bud outer view. D) Patch bud inner view. E) Patch bud inserted and wrapped with PVC budding tape.

budstick with a sideways push rather than by lifting it, otherwise the middle of the bud may be left behind. Immediately transfer the bud patch to the prepared stock and wrap with budding tape. Pull the wrapping tight enough to ensure good contact between the cambial layer of the scion and the rootstock, and to retain moisture, but not so tightly over the eye of the bud as to damage it.

Should the bark of the rootstock be thicker than that of the scion, you will need to pare down the rootstock bark around the scion patch until it is thinner than the inserted scion bark. This ensures that the budding tape does exert pressure on the scion patch.

To get accurate and equal spacing between the top and bottom cuts on stock and scion, you can make a small double bladed tool by clamping two (razor) blades either side of a spacing block.

Variations of this technique include ring budding and circular patch budding. For ring budding, a complete ring of bark is removed from the stock and replaced by a similar sized ring of scion bark that includes a bud. In the circular patch technique, a round or oval patch of bark is cut from the rootstock with a punch and replaced with a bud-bearing patch cut from the scion using the same or a very slightly larger tool.

V-budding

V-budding is an experimental technique that has been used with young seedling citrus rootstocks as an alternative to microbudding. Like microbudding, it has the advantages of rapid callusing because the young

Figure 16. V-budding of citrus. A) Rootstock prepared. B) Bud cut from scion, side and top view. C) Bud inserted and ready for wrapping. D) Bud growth eight weeks after budding.

tissues are actively growing. Also, the young bud wood used is readily available. The V-budding technique has the added advantage that you keep all the leaves on the rootstocks, so there is no check in growth. Disadvantages are that buds and stocks of this size are not easy to manipulate, and young scion wood has a limited storage life.

The technique involves two cuts on the rootstock and scion, making a V shape as follows:

1. Start the first cut from the leaf side of the bud, and extend it about one-quarter of the way through the rootstock at 45 degrees down towards the base. Retain the leaf.
2. Begin the second cut just above the bud, passing behind the bud towards the base of the rootstock to meet the first cut (Figure 16A).
3. Make two similar cuts to remove a bud from the scion.
4. Insert the scion bud in the rootstock, and wrap with narrow budding tape (Figure 16C).

Grafting techniques

Grafting is preferred to budding if budding is unreliable for a certain variety, when the season is unsuitable for budding, or if the bark of the rootstock is too thick or too thin for successful T-budding.

However, grafting has several disadvantages when compared to budding. It takes longer to cut the stock and scion, match the cambial layers, and wrap and cover the graft. Much more scion wood is needed for grafting. Also, the older scion wood may be more woody and harder to cut. On the other hand, older scion wood has a longer storage life.

Scion wood used for grafting may be either dormant or green. Take it from the last 30 cm or so of the most recent mature flush of growth, cutting from healthy shoots of a similar diameter to the available rootstocks. Cut the scion wood into any convenient lengths for storage. Recut it when grafting, into scion pieces about 5 cm long bearing at least two buds.

For deciduous plants, collect scion wood after the stage of deep dormancy and immediately before the first signs of growth in spring. It may be used at once, or stored in sealed plastic bags at about 5°C for use during the following six months. Scion wood remains viable as long as the cambial layer appears green, not brown or black.

Scion wood of evergreen species is usually collected immediately before the spring flush of growth, but may also be collected immediately before subsequent growth flushes. Scion wood of plants from cool, temperate and subtropical regions should remain viable for up to several months if stored in sealed plastic bags at 5°C. As you collect the scion wood, remove all the leaf blades, but retain the petioles.

Actively growing shoots can be used as a source of fresh 'green' scion wood, from spring right through to autumn, for green grafting both deciduous and evergreen species. Cut off and discard the soft, easily bent shoot tip, then take the next 10 cm or so of the shoot as scion wood. For some plants, such as annona, mango, and cashew, you often get a better success rate if you prepare the scion wood some time before taking it, as follows. Cut off and discard the soft shoot tip, then remove the blades from the last two of the remaining leaves, leaving the petioles attached to the shoot, and the deleafed shoot attached to the plant. One to four weeks later, when these petioles have fallen and the buds have begun to swell, the scion wood is ready to use. Green grafting is illustrated for the annona in Figure 29A–D.

Common grafting techniques described here include the splice graft, whip and tongue, wedge graft, bark graft, side graft and approach graft.

Splice or whip graft

With this graft, you must hold the scion in place on the rootstock until you have finished wrapping, so it is a one-person operation. In a commercial situation, grafting is often a streamlined team job, with one person cutting and another tying; in this case you would whip and tongue or cleft graft. The splice is a simple and easy graft to do when the scion diameter and bark thickness matches with the rootstock and where the rootstock is comfortably accessible.

1. Choose a straight part of the rootstock, about the same diameter as the available scion wood. Make a sloping cut, diagonally right through the rootstock shoot, at least three to four times as long as the width of the shoot.

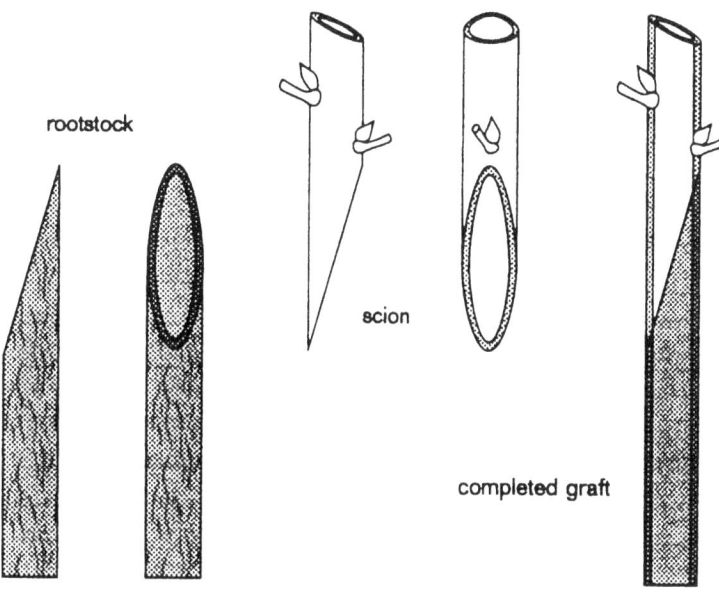

Figure 17. Splice or whip graft.

2. Then make a similar cut across the scion, so that the two cut surfaces are the same size and shape. It is essential that the cut surfaces of the rootstock and scion are as flat as possible to ensure good contact of their cambial regions.
3. Clamp the cut surfaces of scion and rootstock against one another between thumb and finger. Adjust the two parts until the lines of exposed cambial layers at the junction of wood and bark coincide closely. If you cannot match the cambium on both sides because of differing shapes or widths, then match as well as possible on one side and end only. Do not centre a poorly matching scion on the stock, as this will lead to a bad match all around and probable failure.
4. Firmly wrap the graft with budding tape, starting from just below the join and finishing above, in the same fashion as described for T-budding. Try to hold the two pieces fixed in their matched position while wrapping. If they slip, re-adjust as necessary.

Dormant scions need no other protection than a dab of grafting mastic or wound dressing paint on any exposed cuts. However, you must protect green or evergreen scions from drying by maintaining an environment of high humidity around the new grafts until the union is complete. In the humid tropics, shading alone may be sufficient. In other places you should cover the graft in some way, with a sealant of plastic water-based paint, or a special graft sealant, or with plastic film.

You can also continue wrapping with budding tape right up over the scion, in the fashion used to cover a sore finger. To do this, after you have firmly wrapped the graft area, clamp and hold the last turn of tape between the thumb and the scion. Next, take the tape up and over the tip of the scion, and down again to hold beneath the forefinger on the opposite side of the scion. Then, while holding this loop over the tip of the scion, continue wrapping around the scion from base to tip and down again, overlapping the turns carefully. Tie off in the usual way by slipping the end of the tape under the last turn and pulling tight. If you only have a few grafts to do, and no other covering material to hand, this is a good cover.

These days, with the ready availability of various sizes and types of plastic bags at reasonable cost, you would more often cover the completed graft with an upturned small plastic bag, either tied, or even held in place and sealed by the zip-top (Figure 8). If you protect the graft with a plastic bag, you must shade the plants from direct sunlight on hot days, so that the temperature in the bag does not rise to lethal levels. Either hold grafted plants in a shaded area, or cover the plastic bag with a paper bag or some other shade.

In some production systems in dryer regions, instead of covering the scion, the grafted plants are held in a 'fog' house where the humidity is maintained at very high levels with an artificially produced, very fine mist.

Wedge or cleft graft

The wedge or cleft graft is named for the shapes of the cuts made; the cleft in the rootstock, and the wedge on the scion. This seems an easy graft to do, and so is often the initial choice of novice grafters. It is in fact a good choice for species that callus readily.

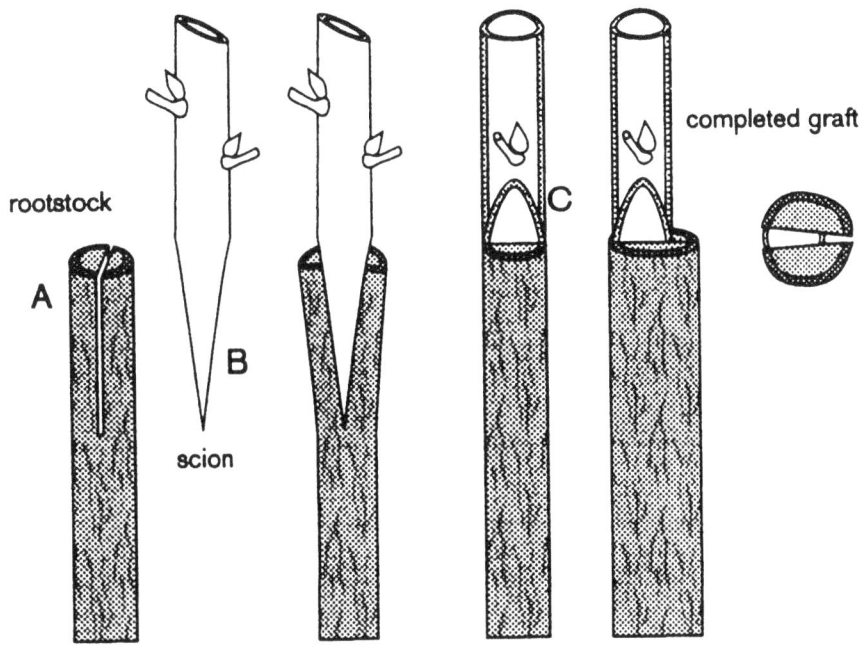

Figure 18. Wedge or cleft graft.

For a wedge or cleft graft, choose a straight part of the rootstock shoot, if possible the same diameter as the scion you have. If in doubt about the match, start where the stock appears a little thinner than the scion. Prune the shoot off square and compare the cut ends of the stock and scion, looking at relative wood diameters rather than the total wood plus bark. You can always cut a little more off the rootstock to get the best match.

Next, make a cut straight down the middle of the rootstock shoot, to a depth of about three to four times the diameter of the scion. Try to cut rather than split the wood (Figure 18A). If the wood tends to split, try a thin bladed knife, and use an angled slicing motion while cutting, rather than forcing the blade squarely down the stock. Some hard wooded plants will usually split, for example, mature pistachio.

Cut the base of the scion to a long wedge (B).

Insert most of the length of the wedge into the slit in the rootstock. Leave exposed the semi-elliptical 'church window' areas of cut surface at the top of the wedge. These areas serve as a source of callus to help heal between the scion and the cut top of the rootstock.

Align the wedge in the cleft so that the cambial areas of scion and stock match. If the scion and stock wood diameters are the same, centre the scion in the cleft. If there is a discrepancy in sizes, match the wood/bark junctions of stock and scion on one side by careful inspection as you slide the wedge into the cleft (C).

Wrap tightly, trying to maintain the position of the scion in the stock. When wrapping, you may find it easier to start at the top of the cleft and then work down. Sometimes, as you start wrapping from the base of the cleft, the scion has an alarming tendency to pop up, especially with sappy, slippery wood.

Wrap carefully with PVC tape, covering the top of the rootstock and the exposed cut areas, and for dormant scions you will only need to seal any cut tip.

Protect a green scion from drying by covering it, as described for the splice graft.

Whip and tongue graft

Matching scion and rootstock

This is simply a splice graft with the addition of a tongue. The tongue helps to hold the graft in position, making it easier to tie. In addition, the scion can be left sitting in position to be tied by the second member of a grafting team. There is increased cambial contact between the rootstock and scion, and the graft is mechanically stronger during at least its first season.

For this graft, make two cuts in the stock as shown in Figure 19A. First, make a long, sloping cut through the rootstock shoot, as for a splice graft. Then, place the knife edge straight across the surface of the first cut, about a third of the way down from the top. Cut into the face of the first

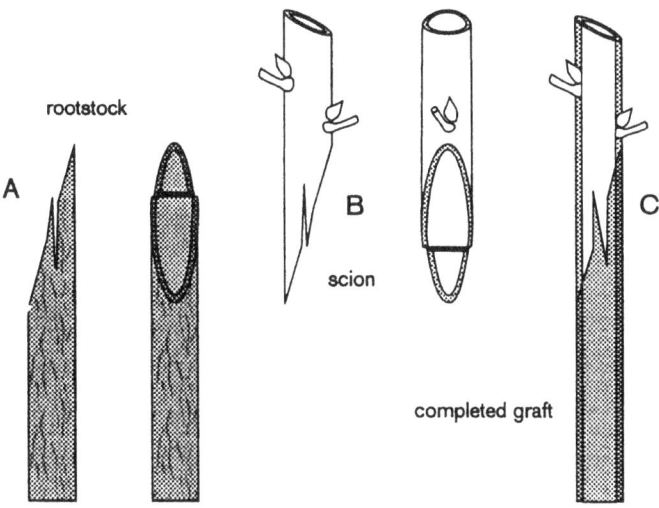

Figure 19. Whip and tongue graft of similar sized scion and rootstock.

cut on a slightly steeper angle, more or less straight down the stock. Make the depth of the tongue about one-third of the length of the first cut.

Cut the scion in a similar way (B). Start the second or tongue cut across and about one-third of the way up the face the first cut, measured from the base of the scion. The correct relative positions of these two tongue cuts in stock and scion may vary according to their angle and depth, which you must learn by experience. Open the tongues a bit with the blade of the knife, or by bending back the tips of stock and scion. Finally, push the two pieces together (C), aligning the cambial areas on at least one side. Wrap and cover.

Smaller scion on a larger rootstock

If the available rootstocks are much larger than the scion wood, you can make a splice graft or a whip and tongue on just one side of the rootstock (Figure 20). Behead the rootstock at a convenient level. Cut a shallow slice from one side of the rootstock shoot, such that the distance between the two lines of wood/bark junction is similar to the wood diameter (less bark) of the scion. Err on the shallower side at first – you can always pare a bit more off.

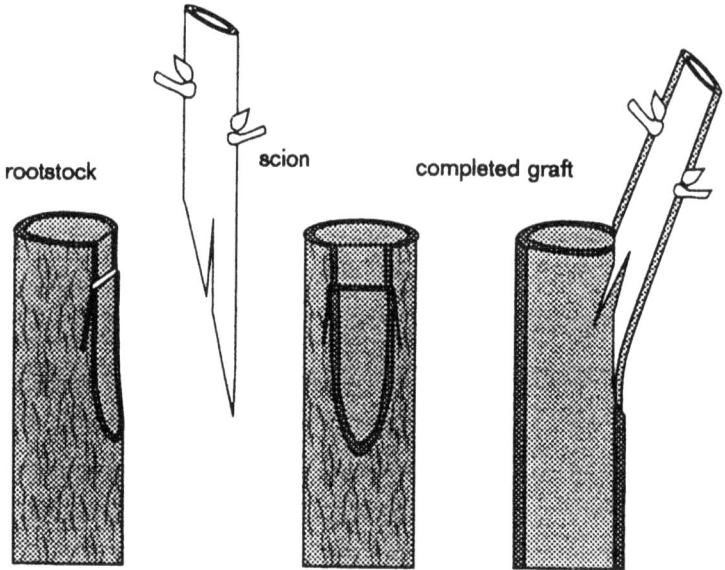

Figure 20. Whip and tongue graft of a smaller scion on a larger rootstock.

Make the tongues as described above for the first type of whip and tongue, possibly starting the tongue on the stock a little less than one-third of the way down. The aim here is to finish with a small part of the cut face of the scion showing above the top of the stock, to aid in healing.

Wrap firmly with budding tape, or use even heavier and wider plastic tape for larger stocks and scions. Seal the top of the rootstock, and cover the graft as necessary.

Bark graft

This is an alternative technique when you have rootstocks much larger than the scion wood. For this graft, the rootstock bark must be lifting easily from the wood.

Slit the bark vertically from the top of the pruned rootstock down for a distance of about four times the diameter of the scion wood. Lift the bark away from the wood on either side of the slit, so that you can easily push the scion down between the bark and the wood.

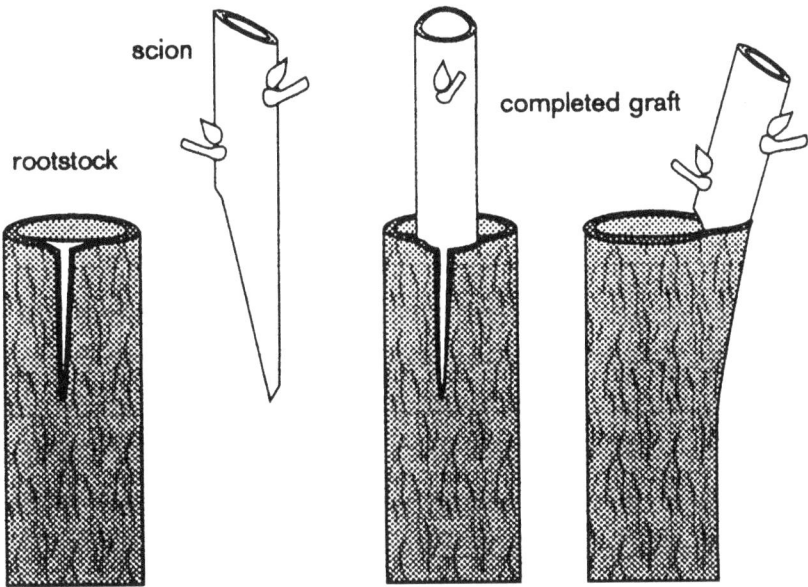

Figure 21. Bark graft.

Cut the scion as for a splice graft, and then remove the end of the thin tapered base with a short cut from the other side (Figure 21).

Insert the scion carefully down beneath the slit bark, with the long, sloping cut surface against the wood of the rootstock. Try not to scrape the scion too heavily against the stock wood, as this might destroy the cambial cells. Stop while there is still a little of the cut face of the scion showing above the top of the rootstock.

Wrap, seal and cover.

Side graft

A side graft (sometimes called side veneer) is another graft used when the rootstock is thicker than the scion. This graft is commonly used for plants such as conifers, which will not tolerate beheading low down on the rootstock where the scion matches for size.

You can add a tongue to this side graft, but it is not essential.

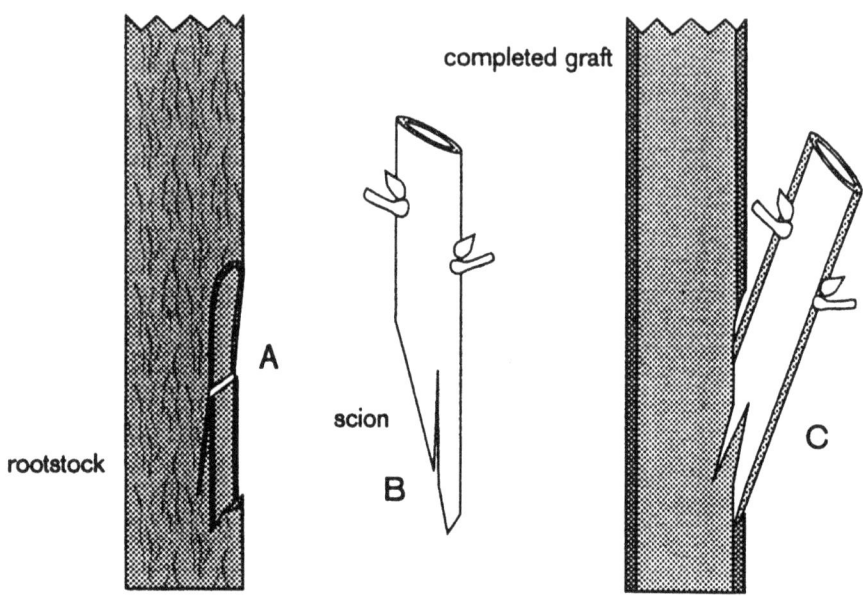

Figure 22. Side graft.

Select a smooth, straight portion of the rootstock shoot.

Towards the bottom of the selected part, make a 3–5 mm deep cut across the shoot, angling it inwards at about 30 degrees from the vertical.

Starting 2–5 cm further up the shoot, make a flat, shallow cut down the side of the rootstock to meet the first cut (Figure 22A). The depth of this second cut should be such that the width of the piece removed from the rootstock is a little more than the diameter of the scion material. Inspect the relative distances from bark to bark on the exposed wood of the stock and on the scion piece, and adjust the depth of the stock cut as needed.

Add a tongue by a third cut about half way down, across the face of the first cut.

Prepare the base of the scion with a long sloping cut on one side, passing not much more than half way through the width of the scion. Compare the lengths of the cut faces of the scion and the rootstock, and then adjust the scion to the correct length with a short, steep cut from the bark side at the base of the scion. Make a tongue (if desired) about one-third of the way up from the basal end of the first cut (B).

Push the scion into place (C), matching the cambium on at least one side.

Wrap, and cover green scions with the corner of a small upturned plastic bag slit half-way down one side, or hold the newly grafted plants in a humid, shaded greenhouse.

Approach graft

True approach grafting involves two whole plants, each with its own shoot and root systems. Since both plants remain intact until the graft appears to have healed successfully, a reasonable success rate is possible even with plants that are otherwise difficult to graft. The scion suffers little (if any) check to its growth with this technique. The longer grafting cuts of an approach graft give the added advantage of extra exposed cambial tissue, and a greater chance of union between rootstock and scion.

Select a long, straight portion of rootstock shoot. Clip off any leaves or shoots that will be in the way.

Cut a strip up to 30 cm long and about two-thirds the width of the scion shoot from the side of the rootstock shoot.

Next, cut a matching strip from the scion, so that on each cut surface, the two lines of exposed cambium at the junction of bark and wood are approximately the same distance apart.

Press the two cut surfaces together and bind in place with grafting tape, matching the cambial regions as well as possible while wrapping. As with all grafting techniques, keep the time from preparation of the parts to completion of wrapping as short as possible.

When the parts have grown together, prune off the head of the rootstock flush above the union, and sever the stem and roots from the scion below the union.

A modification of this technique, called the bottle approach graft, was used during an experimental CSIRO breeding program for avocados. Scions 30 cm or so in length, bearing flower buds, were cut in spring. The

Figure 23. Bottle approach graft of avocado. The base of the scion was kept in water to minimise grafting shock, in order to retain the leaves and flower buds.

buds and some leaves were retained on the scion as shown in Figure 23. Blooms for cross pollination were produced over an extended period from scions stored for up to three months before grafting. After grafting, the base of the scion was freshly cut under water and kept in water, so that the leaves and flower buds would stay alive while the graft took. The rootstock and scion were kept in a humid environment for several weeks after grafting by covering them completely with a large plastic bag.

A type of approach graft in which the scion diameter was much larger than that of the rootstock is illustrated in Figure 24. In this instance, the base of the scion was buried to prevent desiccation. As with the avocado, a long (15 cm) portion of cambium of both the rootstock and scion was

Figure 24. Detached approach style graft of Indian jujube. Growing scion on left, beheaded rootstock on right.

exposed, matched together, and wrapped with budding tape. The rootstock thickened rapidly, and after three months, when the photograph was taken, its diameter was about half that of the scion, whereas initially it was less than a third.

Methods for selected species

Australian native plants

The popularity of Australian native plants blossomed around the 1970s, coincidental with the 'setting up house' (and garden) life stage of the Baby Boomer generation. Unfortunately, Australian plants often were sold and planted indiscriminately, without much thought given to their suitability in terms of climate, soil type, or eventual size. There also seemed to be a general misapprehension that Australian plants, being native, were zero maintenance. Disaster was inevitable, and Australian native plants fell heavily out of favour with the average gardener.

Only since the turn of the 21st century have some newer, more carefully selected varieties started to make their return to general gardening acceptance. One part of the resurgence has been the clonal propagation of new selections, sometimes by grafting.

Eucalypts

Until recently, all ornamental eucalypts were propagated from seed. Not many were propagated by cuttings, unless juvenile plants could be used as a source of cutting material. Enthusiasts have been aware of the possibility of grafting eucalypts for a long time, but grafted plants have only become commercially available since about 2000.

So far, less than thirty selections have been grafted for sale. In all cases, grafting is done to ensure maintenance of the characteristics of the clonal selection (such as flower colour), and incidentally, to induce earlier flowering. Seedling *Eucalyptus ficifolia* takes 7–10 years to flower, while grafted trees can flower from two years after grafting.

Selected scions include some hybrids between the Western Australian flowering gum, *E. ficifolia* (syn. *Corymbia ficifolia*) and *E. (Corymbia) ptychocarpa,* and a few selections of *E. ficifolia* with attractive flower colour and good flower size. There are also some newer hybrids involving either *E. caesia* or *E. leucoxylon* as one of the parents. These are not yet widely available.

Rootstocks are often seedlings of the same species, or related species, for which graft compatibility has been established. The rootstock species should also have demonstrated good horticultural qualities, such as uniformity of seedlings, good health and growth performance over a wide range of garden conditions, and appropriate vigour. In the case of *E. ficifolia* scions, rootstocks are usually other eucalypts from the bloodwood section, such as *E. maculata* or *E. gummifera* seedlings. Seedlings of *E. ficifolia* are rather variable, but have also been used as rootstocks. For scions from the large subgenus *Symphomyrtus*, there is a range of rootstocks possible.

Progress is now being made in the production of cutting-grown rootstocks.

Collect fresh scion material in summer, from healthy, strongly growing plants. It is advisable to collect scion wood from well-grown garden plants that have been kept in vigorous growth by appropriate watering, fertilising and pruning. Choose material from the spring flush that has hardened off, and that is showing some plump vegetative buds. Clip all mature leaves from the scion, and keep in a cool humid place until grafted. Scions are best collected and used on the same day. As eucalypts are not always easy to graft, some people have suggested that you prepare scion wood by prior cincturing, as described for macadamia later in this section. This preparation may increase success with scion material from trees growing in less than good garden conditions.

Cut off the top of the seedling rootstock at a point where it appears somewhat thicker than the scion material. Carefully match the wood diameters of scion and stock, cutting again further down if necessary. Leave at least several pairs of leaves on the rootstock plant.

You could graft by any of the commonly used methods, such as wedge, splice, or whip and tongue. It is also possible to bud.

The side wedge graft illustrated for passionfruit might also be suitable – see section on passionfruit. This would allow retention of most of the leaves on the rootstock, with only the terminal bud removed. Cut the stock back after the graft has taken.

Bind the graft in place with budding tape, and hold the grafted plant in a warm (25°C), area, under 50% shade and high humidity (80% plus). For small lots, cover the scion with a small plastic bag.

Grow the grafted plants on for sale or planting out after one to two years.

Banksias

The large and spectacular flower spikes of the Western Australian species of *Banksia* make this group highly desirable as garden plants. Unfortunately, many Western Australian species have proven unreliable garden subjects in the wetter parts of the eastern states of Australia (Melbourne and Sydney) mainly because of their susceptibility to root diseases. The obvious solution would seem to be grafting onto rootstocks of the hardier Eastern Australian species, such as coastal banksia (*B. marginata*). Amateur growers in this field have done a lot of research, but so far, long-term success has eluded them. Grafts take and grow satisfactorily, even for a number of years, and then die or break off at the graft union. Other grafts

Figure 25. Binding banksia graft with paraffin film.

initially appear successful, as the scion commences growth, but then growth slows and becomes unhealthy, sometimes with excessive callus forming at the union. However, there have been a few long-lived, successful grafts, which gives hope that further research may yield a solution. Clonal propagation of rootstocks will probably be a factor.

Take scion material for normal grafts (or for budding) from just-hardened shoots of current season's growth in summer. Select scion material from healthy, strongly growing, preferably cultivated plants (scion material collected from the wild does not graft readily). Store temporarily in a cool humid place, for example wrapped in just damp newspaper in an insulated icebox. If possible, use scion wood on the same day as cut.

Rootstock plants should be container-grown seedlings or cuttings in excellent condition, growing actively. Choose rootstocks that are a similar diameter to the available scion wood.

A point worth noting for the successful growing of *Banksia* and other members of the plant family *Proteaceae* is their requirement for low levels

Figure 26. Grafted banksia covered with half a plastic drink bottle.

of phosphorus in the growing medium. Use only potting mixes and fertilisers formulated for growing Australian native plants.

Species worth trying as rootstocks are *B. integrifolia*, *B. ericifolia*, *B. serrata*, *B. marginata* and *B. robur*.

Graft by any of the usual methods of budding or grafting. Pay particular attention to the matching of the cambial regions, taking into account variation in bark thickness between stock and scion.

An additional technique sometimes used with banksias is the grafting of very young rootstock seedlings, just above the cotyledons or seed leaves.

Grow seedling rootstocks on until they are showing one to several sets of true leaves. Pot on into tubes, planting as high as possible to enable access to the part of the stem below the cotyledons. Behead the rootstock seedling, cutting across the stem just above the cotyledons with a sterile, single sided razor blade, slicing down the centre of the stem between the cotyledons for about 1 cm.

Take thickness-matched scion material from 2–5 cm shoot tips of healthy garden plants, or from seedlings of the scion species. Clip off or at least reduce the size of the leaves with sharp, clean secateurs.

Lay the shoot tip on a sterile hard surface, for example, a kitchen cutting board. Cut the base of the shoot tip to a long wedge shape with two strokes of the razor blade.

Slide the wedge of the scion base into the cleft in the seedling rootstock.

Bind the scion and stock by wrapping with a strip of paraffin film around the graft. Stretch the strip of paraffin film before use. A further option might be to splice graft, and to hold the rootstock and scion in place with special small grafting clips or pegs (see tomato grafting later in this section).

Keep the grafted plants in a shaded, warm (25°C) and very humid environment until growth recommences in the scion. The modern equivalent of the 'Bell jar' cover recommended in older publications would be a clear plastic 2 litre soft drink bottle cut in half. Place the trimmed bottle over the grafted plant, with the cut end buried in the pot. The 2 litre bottle fits very well inside a standard 140 mm plant pot.

After scion growth has started, gradually harden off the plant.

Hakeas

Hakeas, banksias and grevilleas are all related, being in the plant family *Proteaceae*. Fortunately, grafting success with *Hakea* has been better, with *Hakea salicifolia* the rootstock of choice. The grafting methods discussed above for *Banksia* will also work for *Hakea*, and there should be few problems with incompatibility between the showy Western Australian species and *H. salicifolia*. The reason for grafting is also the same as for *Banksia*. Western Australian species on their own roots can be unreliable in the wetter, more humid parts of the eastern states.

Grevilleas

Grevillea is a large genus of over 350 species, also mainly Western Australian in occurrence. Some of the same problems mentioned under

Figure 27. Bill Lewis with 12-year-old weeping standard grevillea.

banksias beset grevilleas in cultivation in the eastern states of Australia. But you can readily graft grevilleas by any of the methods described for *Banksia*, with a better long-term success rate. One rootstock that has been widely tried is a Queensland species, silky oak (*G. robusta*). Silky oak has been in cultivation for many years, and is tolerant of a wide range of conditions. It is fast growing and vigorous, eventually becoming a large ornamental tree. It grows easily from the freely-produced seed.

Because *G. robusta* can quickly provide a tall rootstock, you can make standard plants by grafting smaller growing species on a big stock, or make weeping standards by grafting on a scion from a prostrate ground cover species.

Other rootstocks include *Grevillea* 'Poorinda Anticipation', *G.* 'Bronze Rambler' and *G. rosmarinifolia* (useful for smaller growing scion varieties). Propagate these three by cutting.

Collect half-hard scion material from garden grown plants, or from potted seedlings.

A couple of the most attractive Western Australian species you could use as scions are *G. dryandroides* and *G. juncifolia*.

Figure 28. Weeping standard grevillea.

Other Australian natives

Eremophila species (emu bushes) are sometimes grafted, for the same reasons as banksias and grevilleas. Scions are taken from species from the dry inland areas of Australia. Rootstocks can be cuttings of *E. maculata*, or cuttings from various species of the closely related *Myoporum* genus, such as *M. insulare* or *M. montanum*.

Many *Darwinia* species are unreliable in cultivation, but can be grafted to the hardy and long-lived *D. citriodora*.

Prostanthera species (mint bushes) are often propagated by cutting, but some are more reliable when grafted on to either hardy *Prostanthera* species or on to the related *Westringia fruticosa* (native rosemary).

Quandong (*Santalum acuminatum*) is cultivated on a small scale for its edible fruit, which is used for cooking. Naturally, growers would like to multiply up superior selections of quandong that have better quality fruit.

Rootstocks are seedling quandong, grown in a pot with a low-growing host plant such as strawberry clover. Quandongs are partial root parasites, and grow better when a host plant is provided.

Top graft seedlings at 2–3 mm diameter, with a wedge or splice graft. Wrap and cover, or hold in a very humid, shaded environment until the scion starts to grow. Healthy rootstocks and quality scion wood will lead to better success.

Several species of Australian native citrus are now in cultivation. Desert lime, *Citrus glauca* (syn. *Eremocitrus glauca*), grows to a columnar large shrub to 4 m x 2 m. It has narrow, grey-green leaves, and small (1–2 cm diameter), thin-skinned yellow fruit. The fruit has a rather strong, aromatic, spicy flavour, and makes excellent jams and chutneys. Wild trees are often very thorny, but an almost thornless variety was released by CSIRO Plant Industry as 'Australian Outback Lime' (PBR, Plant Breeder's Rights granted).

Experience so far indicates that citrange is a good rootstock for native citrus. With good quality scion wood from cultivated trees, you can T-bud as for normal citrus. Splice grafting works better if you collect scion wood from wild stands.

Some hybrids involving other native citrus (finger limes, *C. australasica*) are also available.

Annonas or custard apples

The Annona family includes the cherimoya (*Annona cherimola*), the sweetsop (*A. squamosa*), the soursop (*A. muricata*), the custard apple or bullock's heart (*A. reticulata*), and the atemoya (*A. squamosa x cherimola*), which is the custard apple sold in Australia. For all these species, grafting is the preferred propagation technique. Graft either in spring with grey-green mature wood of the previous season's growth, or green graft during the growing season. With both techniques, use vigorous seedlings as rootstocks.

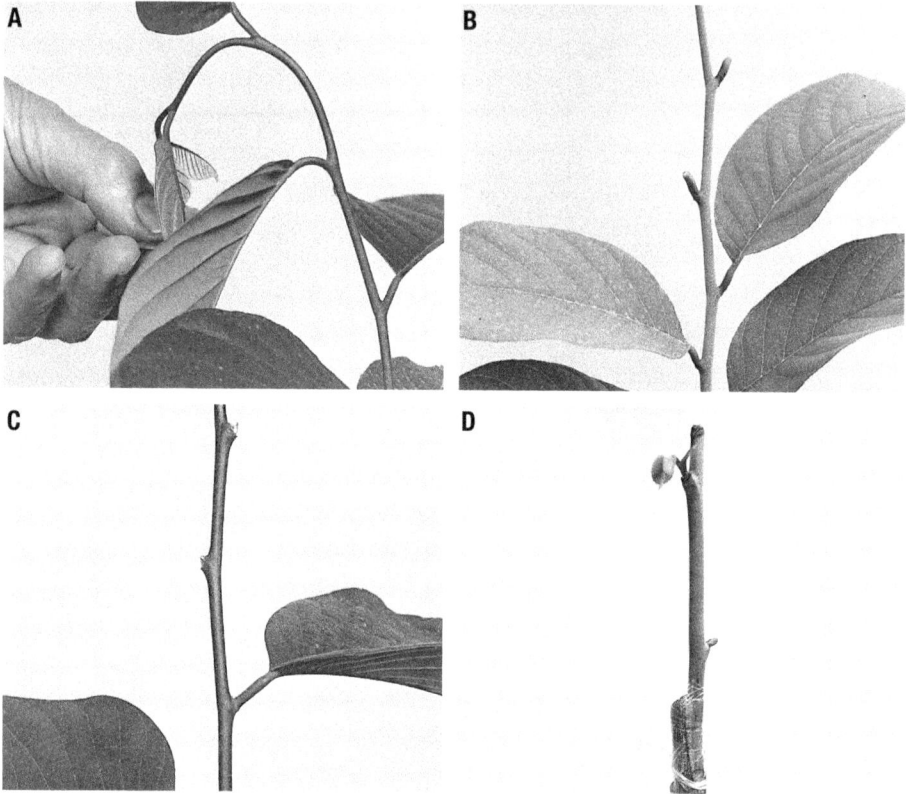

Figure 29. Green grafting of Annona. A) Selection of scion wood – the portion of stem below the bend is used. B) Preparation of graft wood – shoot tip and leaf blades of terminal leaves are removed. C) When the petioles fall, the graft wood is ready to use. D) Scion growth approximately three weeks after grafting by the bark graft technique.

Annona buds develop right inside the hollow petioles, so take care when collecting scion wood and clipping off the leaves. To avoid damaging the buds, retain at least 5 mm of each petiole base.

As annona shoots mature, the bark colour changes from green, to grey-green, to light brown. Grey-green scion wood is best for splice grafting onto one-year-old rootstocks as it matches well to rootstock shoots of a similar age.

Seedling rootstocks three to six months old can be green grafted, using scion wood from young, green shoots of the current season's growth, prepared as previously described (see Figure 29). Graft the rootstock anywhere below the soft shoot tip where the scion and rootstock diameters are about the same, using a whip and tongue or a splice graft.

Older, larger rootstocks can be bark grafted (Figure 29D).

Bud burst of the scion is rapid and may be as soon as two weeks after grafting.

The green grafting technique can be used for a range of other species including jujube (*Ziziphus jujuba*), persimmon (*Diospyros kaki*), longan (*Euphoria longan*) and casimiroa (*Casimiroa edulis*).

Avocado

Grafting avocado plants is generally more successful than budding them. For the best results, collect scion wood just before the spring growth flush.

At this time, which in southern Australia is about the end of July, shoots contain a high level of stored starch, and the stems are stiff and woody. Cut the terminal 5 cm from fully exposed shoots and clip off all the leaf blades, leaving part of each petiole attached to the scion wood (Figure 30).

Figure 30. Avocado shoot selected for scion wood. Pruning cuts are indicated: T) terminal scion wood, K) knuckle. The ring of buds that was the tip of the previous growth flush is indicated by the arrow.

The petioles will fall naturally, either during storage, or when the graft starts to grow. Store the scion wood under refrigeration at 5°C in sealed plastic bags.

Avocado scion wood in refrigerated storage remains viable for four to five months until the initially green cambial layer begins to darken to brown or black.

Alternative sources of scion wood are terminal shoots collected just prior to a later growth flush, or the part of a shoot back from the last growth flush.

This older portion of shoot carries a ring of closely spaced buds, which formed around the growing point of the previous flush when its shoot elongation slowed down. Pieces of graft wood with this ring of buds are termed knuckles and you can cut them at any time (Figure 30, K). However, if you need a lot of scion wood, cutting back to knuckles may strip too many leaves from the scion source trees.

Rootstocks for avocados are generally vigorous seedlings, grown from seed that has been hot water treated at 49°C for 30 minutes to eliminate phytophthora root rot fungus.

Graft the rootstocks when the shoot diameter above about six basal leaves is similar to that of the scion wood. At grafting time, maximum temperatures should be between 20°C and 25°C. Graft using any of the splice, whip and tongue, wedge, bark graft, or side graft techniques.

Wrap the graft union tightly with budding tape. Protect the scion from drying, either by completely wrapping the scion with budding tape or by covering it with a small plastic bag. To raise the humidity, you can trim the two uppermost leaves of the rootstock, and include them in the plastic bag.

When grafting in the open, use a paper bag or some other shading to keep down the temperature inside the plastic bag.

Depending on growing conditions and the vigour of the rootstock, the graft will commence growth after two to six weeks. You should remove

Methods for selected species | 51

Figure 31. A whip and tongue graft of avocado ready for wrapping.

the grafting tape completely after two to three months or else it may strangle the scion growth. Support the growing graft by a stake for at least its first season, especially if it is exposed to strong winds.

As part of a CSIRO avocado breeding program, a technique was developed to graft tiny avocado shoots from embryos germinated in tissue culture.

Figure 32. Avocado. A) Tissue-cultured embryo shoot bark grafted to 12-month-old seedling. B) Six weeks after grafting.

Embryos at least six weeks old were rescued from prematurely fallen fruits set from hand-pollinated flowers, and then successfully grown in tissue culture.

When the shoots were about the size of a match, they were grafted as follows. Young, vigorously growing rootstocks about 30 cm high were pruned off just below the soft growing tip. The pruned top of the rootstock stem was pared off with a sharp knife to speed healing. Two parallel, vertical cuts were made through the bark at the top of the stock, separated by the width of the scion shoot and about two-thirds its length. The bark between them was peeled back and cut off to leave a short flap at the base.

The scion shoot was then removed from the tissue culture tube and washed free of agar medium. Using a small surgical scalpel, a long, sloping cut was made down one side of the shoot, then a short, steep cut from the opposite side at the base. The base of the scion was inserted under the prepared bark flap, with the long cut surface held against the exposed wood of the rootstock while it was wrapped with narrow budding tape (see Figure 32).

Part of the uppermost two leaves of the rootstock was then inserted in an inverted plastic bag to provide a humid atmosphere and some shading for the graft. The bag was removed after three to four weeks when the graft started to grow.

Cacti

Many members of the cactus family (*Cactaceae*) grow readily from seed, or in the case of selected clones, from cutting. However, some selections are weak, or slow growing, or unable to survive unless grafted.

Examples in this last category are the brightly coloured red or yellow 'Ruby Ball' or 'Moon Cactus' (*Gymnocalycium* species) that lack chlorophyll. Seedlings of this type cannot photosynthesise, and will die after a month or so unless grafted onto a rootstock with a length of green stem.

Slow growing varieties are also grafted to improve vigour and growth rate. These include the many cristate (crested) selections of *Mammillaria* species and various crested, variegated or monstrous species in other genera. The faster growth consequently allows economies of scale in commercial production, and sometimes earlier flowering as well.

The graft used for cactus is technically interesting, as it is about as plain as a graft could be. Usually called a flat or butt graft, it consists of simply making a flat transverse cut on both stock and scion, then holding them together with the cambial areas matching.

There is quite a range of different rootstocks used in the grafting of cacti. Unusually, the choice of rootstock in the *Cactaceae* is not limited by considerations of compatibility between different genera. Most combinations will graft together satisfactorily, so choice of stock is based on other factors, such as the climate of the location and the requirements of the grower.

For example, for rapid large-scale commercial production, 'Moon Cactus' is grafted onto the easily propagated and fast growing *Hylocereus trigonus*. A drawback of this easily produced combination is that as the plant ages, the green epidermis of the *H. trigonus* rootstock gradually changes to a corky, light brown bark, thus losing its ability to photosynthesise. Eventually the plant will starve to death, unless re-grafted onto a fresh green-stemmed stock. In addition, *H. trigonus* does not tolerate cold conditions (below 10°C). A longer-term solution is to use a rootstock that does not develop a corky bark, such as *Cereus peruvianus*, which is also more cold tolerant.

Some other widely used rootstocks are as follows:

Myrtle cactus (*Myrtillocactus geometrizans*) is vigorous, propagates easily from cutting, and is relatively trouble-free in temperate climates.

Pereskiopsis spathulata propagates very easily from cutting, and is useful as a stock for very small scions. It is widely used in the USA as an understock for *Lophophora* species, the source of the hallucinogen peyote. Stems of *P. spathulata* are relatively small in diameter (less than 10 mm) and bear true leaves, a rarity in the cactus family. Scions grafted to this species grow quickly and flower early, but the combination may not be long lived, partly due to scion overgrowth.

Trichocereus spachianus (torch cactus) is a many-ribbed prickly columnar cactus that makes a hardy and dependable rootstock.

Opuntia compressa is a moderate growing prickly pear-type cactus that is cold-tolerant and long-lived. Its paddle-shaped sections propagate easily by cutting.

To make a graft

Select a rootstock plant that is growing vigorously, but one that has not been watered in the last few days. Irrigation can lead to excessive sap flow from the cut surface of the stock. Match roughly for size with your proposed scion.

To make your grafting cuts you will need a very sharp knife, or a razor blade, or disposable blade knife. The blade should be sterilised with 70% methylated spirits/water between cuts.

To hold the scion in place once the graft is completed, you can use light string, or thin rubber bands of an appropriate size. If the scion is very small, you may not need any binding.

Behead the rootstock with a square cut, that is, at right angles to the axis of the stem. A ring of vascular bundles should be visible towards the centre of the cut stem. These vascular bundles are the site of the cambial regions, where callus will grow to unite the graft.

Chamfer or bevel the edges of this cut. Hold the knife at 45 degrees between horizontal and vertical, and cut off a narrow piece all around the edge of the stem (see Figure 33). This is an essential step, which prevents distortion of the cut surface, as the skin and flesh dry and shrink at different rates. If not chamfered, the cut top will become concave as it dries, lifting the centre of the scion away from contact with the stock. Use a sharp, serrated kitchen knife to make these initial beheading and bevel cuts, as the skin may be tough and hard to cut.

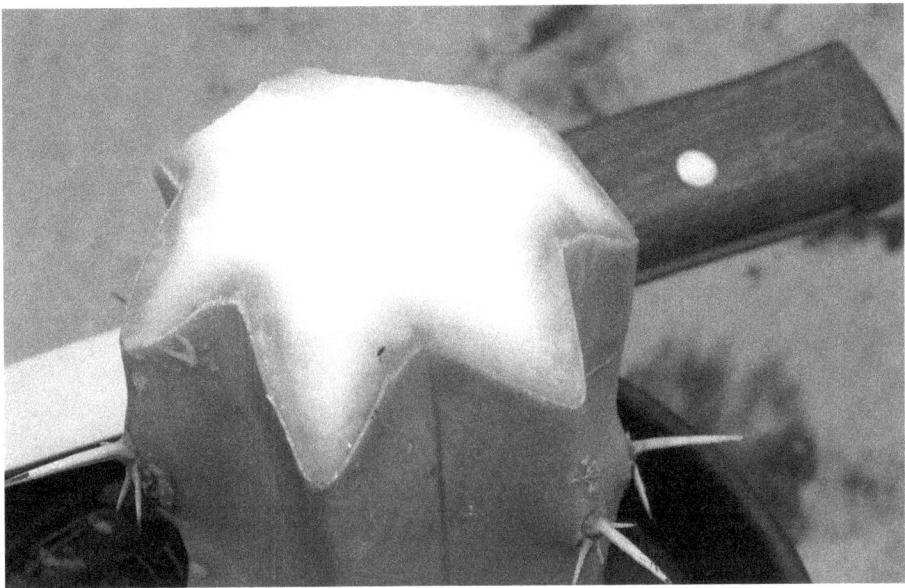

Figure 33. A Myrtillocactus stock prepared for grafting.

Slice off the base of the scion piece. On inspection, you will see the ring of vascular bundles on the cut surface of the scion. If the scion has a thick, tough skin, bevel the edges of the scion as well.

Re-cut a thin slice from the top of the stock, and discard the slice.

Place the scion on the stock. Apply gentle finger pressure (wear a glove or use a tool for prickly species) and slide the scion around a little to remove any air bubbles. Then locate the scion on the stock so that at least some parts of the two rings of vascular bundles coincide. A perfect match may not be possible, but ensure that the two rings at least intersect in a couple of places. Do not centre a small scion ring completely inside a larger stock ring.

To hold the scion in its chosen spot, stretch a rubber band over the scion and right under the bottom of the pot. Apply a second rubber band at right angles to the first.

Choose the diameter and thickness of the rubber bands to hold the scion firmly in place, but not so tight as to squash it or pop it right off the top of the stock!

Hold the completed graft in a shaded, dry area at about 25°C. New growth in two to four weeks indicates the success of your graft. The success rate in grafting cactus is usually good, 80% or higher, even for inexperienced grafters.

Cashew

Under good growing conditions, you can propagate cashew by any of the grafting or budding techniques described in this book. All methods are equally satisfactory.

The main ingredients for success are an actively growing rootstock, well supplied with nutrients and water, temperatures between 20°C minimum and 33°C maximum, and prepared scion wood of a suitable size, at least two months old.

Prepare the scion wood by disblading the leaves one or two weeks beforehand, as described for green grafting. When the petioles fall and

Figure 34. A) Cashew scion wood prepared by removing leaf blades from a mature terminal. B) The scion wood is ready to use after the petioles have fallen, and the buds have started to swell. C) Whip and tongue graft of cashew on 12-month-old seedling, three months after grafting.

the buds start to swell, the scion is ready to use (Figure 34). With this technique, the only fresh wound is the grafting cut. This is particularly important for cashew, because freshly cut surfaces exude a sticky sap that provides an ideal medium for fungal growth.

Actively growing two- to three-month-old seedlings are the preferred rootstocks, grafted by the cleft or whip and tongue methods. If the diameter of the rootstock is larger than that of the scion, use a bark graft.

Under conditions of low humidity, prevent the scion from drying by enclosing it, together with the upper two leaves of the rootstock, in an inverted plastic bag.

Should any fungus develop on exposed cut surfaces or leaf scars, brush or spray the scion with a general fungicide.

You can remove the plastic bag after about one month when the graft starts to grow. Remove the budding tape after about three months.

Young cashew seedlings less than two months old can be patch budded using the following technique:

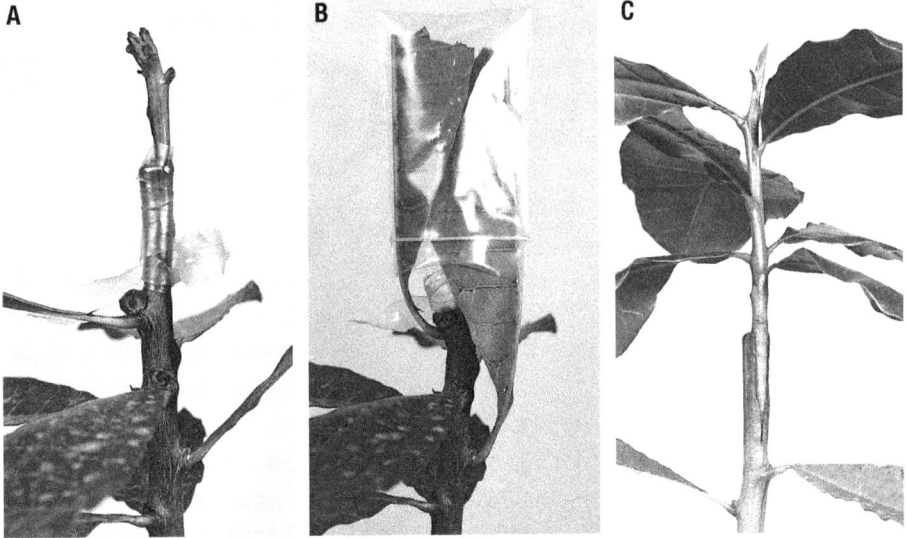

Figure 35. A) Cashew bark graft completed. B) High humidity is maintained around the graft by covering the scion and the top two leaves of the rootstock with a plastic bag. C) Developing graft – three months after grafting.

1. Prepare scion buds by removing the immature terminal shoot tip and the leaf blades of several buds on shoots with a diameter approximately that of the seedlings. When the buds begin to swell and have developed to approximately 5 mm long with a distinct point, they are ready for budding.
2. Prepare the seedling rootstock by removing the shoot tip above the initial rosette of leaves (Figure 36).
3. Patch bud on the smooth, uninterrupted stem below the rosette. A patch surrounding the bud of approximately 6 mm width and 20–25 mm length is sufficient. Split 12 mm wide budding tape into strips 6 mm wide to use with the small stems and buds.
4. After about four weeks, when the patch bud begins to grow, remove the leaf directly above it on the rootstock. Pinch out any rootstock shoots as they start to grow.
5. Remove the tape when the first leaves on the scion are mature (see Figure 36).

This technique has the following advantages: all the available scion wood buds can be used, sequentially down the stem as they develop; large

Figure 36. A) Cashew seedling approximately eight weeks after sowing. Prepared for patch budding by removing the shoot tip and lateral growth. B) Cashew seedling, approximately eight weeks after sowing, patch budded and protected from drying by a plastic bag. C) Scion shoot development (indicated by arrow) four weeks after patch budding.

numbers of plants can be propagated in a limited space; and trees can be budded and field planted before they become pot bound.

Citrus

Citrus plants are almost universally propagated by T-budding, either inverted or normal.

You can collect citrus bud wood at any time when a mature flush is present on the bud wood source tree. The most satisfactory bud wood is the centre third of shoots, with rounded rather than angular wood, carrying plump, well-matured buds (Figure 37). At times when the trees are undergoing a growth flush, such as early spring, it may be difficult to obtain bud wood with sufficient maturity. At these times, semi-mature buds from younger, angular stems may be used, but for these, a micro T-bud is the preferred grafting method. Bud wood may be stored for up to three months at 5°C in a sealed plastic bag.

Because citrus is an important commercial crop, a wide range of different rootstock types has been selected to suit various situations, such as tolerance of wet soils, root rot fungi, and citrus nematodes (citranges and *Poncirus trifoliata*); sandy soils (sweet orange); and salty or alkaline soils (Cleopatra mandarin). Propagate citrus rootstocks from seed of the appropriate rootstock variety. Several seedlings grow from each seed. These can be separated, and all of them grown on for later use.

Seedlings grown from seed of edible citrus may not make satisfactory rootstocks.

When field grown, seedling rootstocks take about a year to grow to a size suitable for budding, which is about pencil thick at budding height. This time span may be reduced by half under glasshouse conditions. Rootstocks must be in active growth at the time of budding.

Stages of normal or upright T-budding of citrus are illustrated in Figure 37. Most commercially produced citrus trees are budded by the inverted T method.

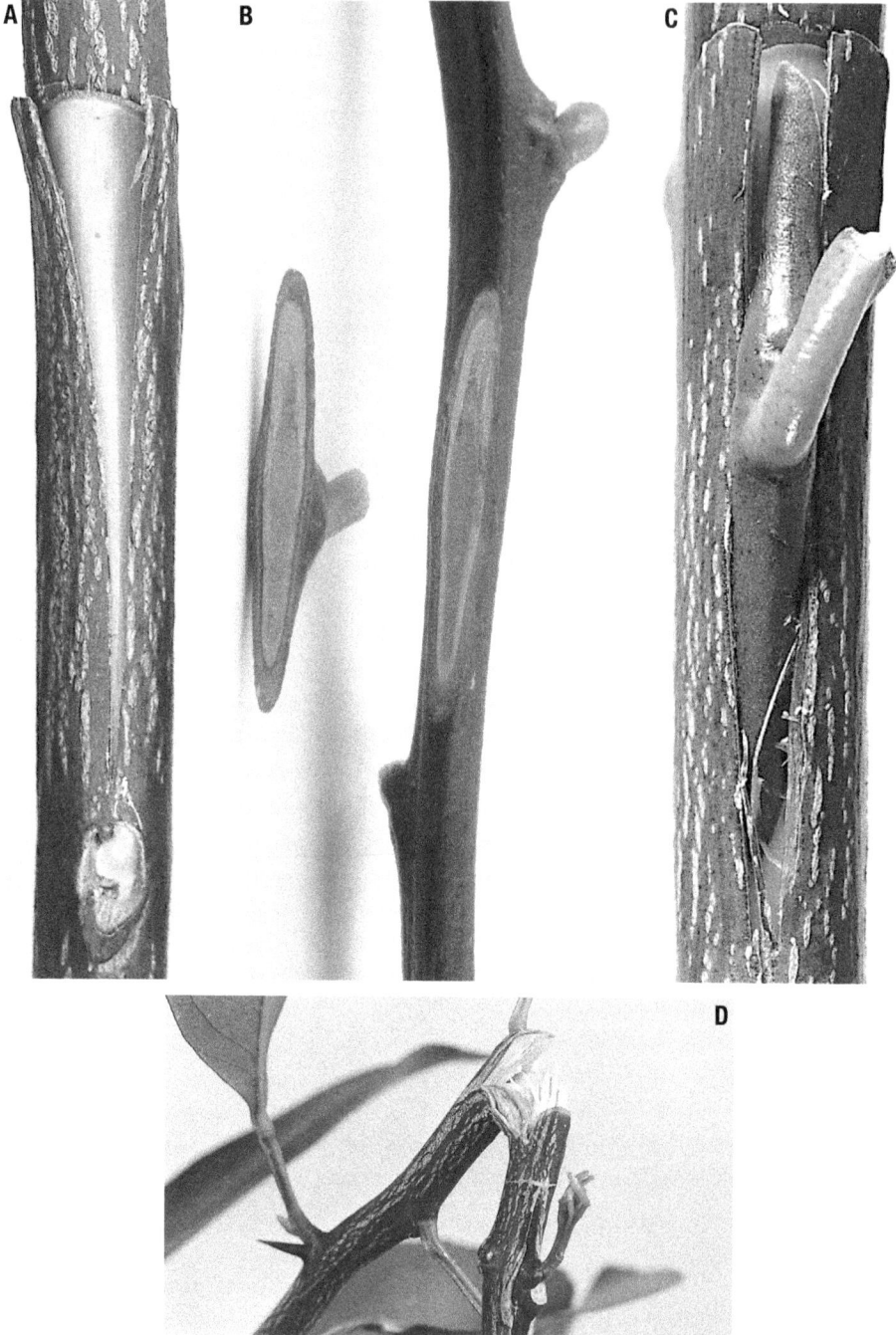

Figure 37. T-budding of citrus. A) Rootstock prepared. B) Bud cut from scion – top and side view. C) Bud prepared for wrapping. D) Bud growth eight weeks after budding.

Best results are achieved by T-budding between August and November or February and April.

To encourage the scion bud to grow, remove the upper third of the rootstock at the time of budding. About three weeks later, either prune off the remaining rootstock shoot above the bud or cut it part way through 5 cm above the bud. Bend over the terminal portion away from the bud. Photosynthesis from the bent-over portion of the rootstock is believed to help maintain the roots in a healthy state until enough scion leaves have developed.

If immature buds are used, they may remain dormant after budding for several months and eventually burst during the following spring or autumn growth flush.

Budding is not very successful for West Indian limes and some mandarin varieties, but they can be grafted. Use a short piece of mature stem carrying one or two buds as a scion for either a side graft or a bark graft.

Very young citrus rootstocks may be microbudded or V-budded. These techniques require a very sharp knife or scalpel and a high level of dexterity, but under good growing conditions will produce rapid results.

Conifers

Conifers, by virtue of their strong architectural forms, have long held a popular place in gardens, especially in the northern hemisphere.

Some of the common species and cultivars can be propagated easily by cuttings, for example pencil pines (*Cupressus sempervirens*), but many of the more desirable and ornamental types must be grafted, as the strike rate for cuttings is too low.

Sometimes the reason for grafting is to achieve a particular form, such as the 'standard' conifers with a bare stem of rootstock of up to a metre in height. These can be ball-on-stick standard conifers (e.g. *Juniperus squamata* 'Blue Star'), or layered standards (e.g. *Cupressus macrocarpa*

'Greenstead Magnificent'), or weeping standards (e.g. *Juniperus taxifolia* var. *lutchuensis*).

Rootstocks for conifers are usually seedlings (or occasionally cuttings) of the same or closely related species. The decision whether to grow from seed or cutting mainly depends on the availability of seed and the ease of striking cuttings. For example, here are some combinations for grafting a few of the more popular conifers.

Cupressus selections are usually grafted onto cutting grown plants of 'Castlewellan Gold' (×*Cupressocyparis leylandii*), a vigorous and fast-growing hybrid that strikes readily from cuttings and grafts easily.

Pinus selections are grafted according to their needle cluster number onto seedlings of *P. thunbergii*, or *P. sylvestris* (two needle), *P. radiata* (three needle) or *P. wallichiana* (five needle).

Picea varieties are usually grafted onto seedling *Picea abies*, or sometimes cuttings of *Picea sitchensis*.

Cedrus selections are grafted onto seedling *C. deodara*.

Producing a grafted conifer

For many conifer rootstocks, you can collect fresh seed in autumn and sow in spring with no pre-treatment.

Seed of some species may germinate more reliably with pre-treatment, as follows:

Sow fresh seed in trays in autumn, in a location where it will be exposed to natural winter break of dormancy. Alternatively, store imbibed seed in a damp medium in a refrigerator (or cool store) at the temperature and duration recommended for the species. This is called cold stratification, and tables or information on the requirements of various species are readily available. Sow stratified seed in spring.

Alternatively, if propagating the rootstocks by cutting, take 10 cm cuttings from current season's growth in late summer. Strip leaves from lower

Figure 38. Conifer side veneer graft.

two-third of cuttings, wound and dip base of cutting in a rooting hormone such as indole butyric acid (IBA). Stick cuttings in trays of open propagating mix and hold in a mist house on benches with bottom heat.

Grow on seedlings or struck cuttings in suitable containers (e.g. 140–150 mm pots) until they are large enough to graft. This is normally about pencil thickness at 50–100 mm above soil level. The time taken to reach this size varies with the vigour of the variety, from 12 months for 'Castlewellan Gold' and up to three years for *Picea* species.

Graft *Cupressus, Juniperus* and *Thuja* species in the summer. Graft by the side graft method (side veneer). Match scion to rootstock carefully for size, trying to achieve coincidence of the cambial areas on both sides of the grafting cut. If you are unsure, make initial grafting cuts conservative – you can always pare off a bit more from the scion, or cut a little deeper on the rootstock. Close matching is particularly important when grafting conifers, since callus formation can be relatively slow.

Bind scion firmly in place with 12 mm-wide PVC budding tape. Seal any cut ends on the scion, and around the graft union, using an appropriate commercial graft sealant.

Carry out the grafting in a shaded humid area, such as a plastic house, and hold the newly grafted plants under these conditions until the scion has shot and matured its first flush of growth, six to eight weeks later.

Leave the grafting tape in place for as long as possible, until it begins to constrict the stem growth. This may be up to six months after grafting.

Graft varieties of *Pinus*, *Picea*, *Abies* and *Cedrus* during the second or third winter after rootstock propagation, by the same method mentioned above.

Taxodium and *Ginkgo* are deciduous genera of conifers. Graft selected varieties of these genera during winter, using a top wedge graft.

Figure 39. Pegged conifer graft in fog house. Grafted conifers are generally ready for sale some two years after grafting.

Tall standard plants of *Cupressus macrocarpa* 'Greenstead Magnificent' are produced by grafting scions onto the top of tall, straight rootstocks of 'Castlewellan Gold'. Figure 39 shows this combination being callused in a fog house, with only a grafting peg and no binding or sealant around the graft.

General deciduous ornamentals

Australia has few native deciduous species, but our gardeners enthusiastically plant many of the well-known deciduous ornamentals originating in northern hemisphere countries. It is a remarkable fact that excepting conifers, nearly all the grafted ornamental plants in gardens are deciduous. This is probably because horticulture developed much earlier and to a greater degree of sophistication in the cold climates cradling the northern hemisphere civilisations.

Most of these plants are chip budded in summer with fresh current season's scion wood that has been refrigerated for 24 hours to harden up slightly.

The eye of the bud is not covered when binding with budding tape.

Popular plants include the following:

Maples (*Acer* species)

Side graft or bud the highly ornamental Japanese maple (A. *palmatum*) onto seedlings of the species.

Red maple varieties (A. *rubrum* and hybrids) are relatively recent introductions for Australian gardens. These are good large shade trees with exceptional autumn foliage. Bud onto stocks of seedling A. *rubrum*. These newer red maple varieties may also be grown by cuttings.

Ashes (*Fraxinus*)

Bud claret ash and golden ash onto seedling *F. americana*. Desert ash (*F. oxycarpa*) was previously used as a rootstock for these ashes, but tends to produce numerous suckers (especially under golden ash) and to overgrow the scion.

Birches (*Betula*)

Cultivars include weeping, cut leaf and coloured leaf types.

Use seedling *B. pendula* as rootstocks. This seed requires cold, damp storage (stratification) before planting.

To make weeping standards, wedge or splice graft weeping birch scions onto tall straight three-year-old stocks. Grow on for another year before planting out, to develop a reasonable sized head.

Ornamental cherries (*Prunus serrulata* selections)

Bud ornamental cherries such as 'Mt Fuji' and various other selections of *Prunus serrulata* onto 'Mazzard' (*P. avium*) seedlings, or any other commonly used cherry rootstock.

To produce tall standard or standard weeping cherries, grow on Mazzard seedlings until two years old. Pick out for straightness of trunk, and top wedge graft. Grow on for another year, or until well-sized.

Figure 40. Field grown one-year-old cherry grafts on stocks grafted at two years of age.

Malus species and hybrids

Bud flowering crab apples (*Malus* species and hybrids) onto seedling Granny Smith apple, or onto one of the clonal 'East Malling' apple rootstocks. Ornamental crab apple varieties are incompatible with 'Northern Spy', a commonly used rootstock for edible apples.

Ornamental pears (*Pyrus calleryana*)

There is now quite a range of ornamental pears (*Pyrus calleryana*) so you can choose scion varieties for desired height and width. 'Capital' is columnar, while 'Chanticleer' (also known as 'Cleveland Select') is narrow conical. 'Bradford' is a larger, broader tree.

Chip bud in summer onto two-year-old *Pyrus calleryana* D6 seedlings.

Mop Top standards

The scion for this tall globose standard is a semi-dwarf, thornless selection of *Robinia pseudoacacia*, usually referred to as 'Umbraculifera', or sometimes called 'Bessoniana'. For rootstocks, grow on seedlings of *R. pseudoacacia* until they are two years old, and about 1.8 m high. Disbud and de-thorn. Cleft graft in summer with two scions on each stock, one each side of the cleft.

This rootstock occasionally suckers if the roots are disturbed in cultivating the soil nearby. Death of the scion or removal of the tree will induce prolific suckering.

Honey locust

Honey locust (*Gleditsia triacanthos*) as it naturally occurs is very thorny on the trunk and branches, so cultivated selections are from the thornless variety *G. triacanthos* f. *inermis*. These selections include a purple foliaged variety, and several with golden young shoots. The trees are ornamental but open in structure, casting a light shade. Bud onto seedling *G. triacanthos* f. *inermis*.

Elms (*Ulmus* species)

Elms (*Ulmus* species). Bud the golden elm and silver elm onto seedling wych elm (*U. glabra*). Bud newer selections of Chinese elm (*U. parvifolia*) onto seedling Chinese elm.

Grape

Grapevines (*Vitis vinifera*) can be propagated easily by cuttings, and vines grown from cuttings are satisfactory in many situations. However, where there are problems with poor soils, salinity, nematodes or phylloxera, grafted vines are more productive.

In large commercial nurseries, grapevines are bench-grafted in winter, using grafting machines. Smaller operators and individual farmers do their propagation in the field or nursery during spring and summer, grafting or chip budding by hand. Grapevines can be T-budded, but usually only when using green scion wood. Any of the grafting techniques described earlier can be employed.

Rootstocks have been bred mainly from related Native American species of *Vitis*, such as *V. rupestris*, *riparia* and *berlandieri*. There are selections to withstand salinity and nematodes (e.g. 'Ramsey') and for tolerance of phylloxera (e.g. 'S04', '5BB' and '140 Ruggieri'). When choosing a rootstock, you should seek some local advice.

Take dormant cuttings of rootstock varieties during winter. Disinfect and cold store in sealed plastic bags for spring planting.

Collect dormant scion wood in late autumn as soon as the vines have lost their leaves, and store it in sealed plastic bags at 4°C. As the growing season progresses, you can bud or graft with dormant scion wood, either in early spring into last year's wood, or in summer into the current season's growth.

Another source of scions during the growing season is 'green' scion wood from semi-mature current season's canes, cut and budded soon afterwards into rootstock growth of a similar age.

If you are chip budding large numbers of vines with dormant scion wood, you can cut the buds beforehand and store them in cold water, picking them out to match for size as you bud. When wrapping chip buds on grapevines, leave the eye of the bud exposed. About a fortnight after budding, cut back the rootstock shoot to one leaf above the inserted bud. The scion should start to grow two to three weeks later. Support the new scion shoot by a string or stake, and regularly rub out any rootstock buds that burst.

Micro-grafting is a technique that enables a rapid build-up of stock when propagating material is in short supply. Small green wood of both stock and scion are machine V-grafted.

Cut current season's canes from vigorously growing rootstock mother vines. Prepare rootstock cuttings, each bearing one central node retaining a leaf blade clipped to about half its original area. Leave 5 cm or more of internodal cane above and below the leaf.

Cut scions in a similar fashion, but with only 3–5 cm of internode below the retained leaf, and a short piece of internode above.

Insert the top of the rootstock and the base of the scion into a V-grafting machine. In one action, the machine cuts a wedge on the base of the scion, cleaves and opens the tip of the rootstock and fits the two parts together.

Bind with paraffin film, or with budding tape, or with thin adhesive polypropylene tape, such as used for bundling vegetables. This tape is coated on one side with an adhesive that sticks reasonably well to the uncoated side of the tape, but sticks poorly to other surfaces such as vegetable stems. As the grafted plant grows, the vegetable bundling tape falls off.

Dip the base of the rootstock in rooting hormone powder, and stick in a suitable small container of growth medium.

Hold the grafted cuttings in a controlled climate room at very high humidity and reduced light for up to one month, until the graft union has healed and the rootstock has taken root.

Harden gradually to outside conditions.

Figure 41. Micro-grafted grape vine.

You can re-work established grapevines in the field to a new variety by sawing off the trunk and wedge grafting in spring with stored dormant scion wood. Seal the cut surface of the trunk and the cleft with graft sealant.

Macadamia

Collect scion wood for macadamia in July, just before the start of the spring flush of growth. At other times of the year, prepare scion wood by removing a ring of bark 1–2 cm wide from around the base of an exposed shoot or branch (cincturing). The interruption of the conducting tissues of the bark forces starch to accumulate in the wood beyond the cincture. Cincturing considerably improves the success rate of grafting macadamia.

Collect the scion wood four to six weeks after cincturing, when callus is obviously developing on the far side of the cincture cut. Select pieces with a similar diameter to that of the rootstocks and with two rings of buds.

Cut scion wood as required, although you can store it under refrigeration at 5°C for several weeks.

Grow seedling rootstocks from macadamia nuts in shell, planted while fresh, since viability decreases rapidly in storage. Good growing conditions and rootstocks in active growth are particularly important to achieve a reasonable success rate when grafting macadamias. Fertilise only with low phosphate 'native plant' fertilisers, because higher levels of phosphorus can injure macadamia plants.

Closely match for size rootstocks and scions of about pencil thickness and splice graft. When matching, take note of the internal wood diameters rather than the total wood plus bark thickness. Scion bark is often thicker, particularly with cinctured scions.

Because macadamia wood is rather hard, it is difficult to make flat grafting cuts, especially with larger diameter wood. If the cuts are not flat, this leads to gaps between the two parts of the graft. It is difficult to close these gaps when wrapping, due to the inflexibility of the hard wood. Some operators use a small cabinetmaker's wood plane to finish the stock and scion to a flat surface for good matching. Wrap the graft as tightly as possible, and cover the scion.

Macadamia can also be propagated by chip budding and seed grafting. In seed grafting, the emerging shoot of a germinating seed is cut off flush with the shell. A small scion of prepared wood is cut to a sharp wedge and forced into the cut surface of the shoot.

Mango

Mango, like avocado, is often more successfully grafted rather than budded.

Seedlings of the varieties 'Common' and 'Kensington' are used as rootstocks in Australia. 'Kensington' seedlings are more vigorous and will give a larger tree. Before sowing, immerse the seeds in hot water at 50°C for 20 minutes to reduce fungal and insect problems. After heat treatment, remove the tough fibrous outer covering from the seeds, and plant on edge with the convex side up. The above-mentioned mango varieties usually produce several seedlings from each seed, but it is common practice to discard all but the strongest.

Grafting young, actively growing seedlings with freshly collected scion wood that is about to flush gives the best success rate. At other times, scion wood may be prepared by removing the leaf blades as described for green grafting (Figure 42A). After approximately two weeks, the petioles fall and the buds swell (Figure 42B). At this stage, the scion wood is ready for use.

Scion wood prepared in this way has the advantage of the absence of sticky sap, which exudes from freshly cut petioles. Mango scion wood remains viable in storage for only one or two weeks so use it as soon as possible after collection.

Mango can be grafted using any of the techniques described in this book. Retain at least eight mature leaves on the rootstock to ensure that the roots remain healthy.

Figure 42. A) Mango scion wood prepared by removing leaf blades from a mature terminal shoot. B) Scion wood is ready for use when petioles have abscised and buds have started to swell. C) Mango successfully grafted by the whip and tongue method.

Under dry conditions, you should cover the completed graft with a plastic bag to keep the scion alive until it has united with the stock. Trim the two upper leaves of the rootstock to size and include in the plastic bag to maintain high humidity. If the graft is exposed to direct sun, you may have to cover the plastic bag with a paper bag, or otherwise shade it to prevent overheating.

Grafts normally begin to grow two to three weeks after grafting. At that stage, remove the plastic bag. Remove the budding tape after four to six weeks, when the new graft has produced some fully formed, mature leaves.

Passionfruit (black passionfruit, *Passiflora edulis*)

Fresh passionfruit seed germinates readily (some other species of *Passiflora* are environmental weeds in Australia and elsewhere). Under suitable conditions, the vines grow strongly, and bear fruit in their second season. As passionfruit is not a highly selected or improved species, the fruit from seedlings is of acceptable quality. Thus, seedling passionfruit are often available at relatively cheap prices.

However, seedlings can be short lived and may never even reach fruiting age in poorly drained situations that favour the root disease, fusarium wilt. So a large proportion of the passionfruit vines offered for sale in Australia are grafted onto fusarium-resistant stock. Nematodes can also be a problem. Grafted plants may have a productive life span of three or four years, after which even they succumb to fusarium, nematodes, or passionfruit woodiness virus, which is widespread and readily transmitted by sucking insects.

Several different *Passiflora* species are used as rootstocks. In the commercial growing areas of northern New South Wales and southern Queensland, selection is continuing for seedling lines of the yellow-fruited passionfruit most resistant to fusarium. In cooler southern areas, for home garden use, banana passionfruit (*P. mollissima*) and blue-flowered passionfruit (*P. caerulea*) are more cold-tolerant, although the latter tends to sucker if the roots are disturbed. The small New Zealand passionfruit growing industry relies entirely on seedling black varieties, as the climate is too cold for yellow-fruited passionfruit, and trials have shown no advantage in using the other two rootstock species mentioned above.

There has been some work to produce improved scion strains of black passionfruit. The Australian Passionfruit Industry Association has its own breeding program, mainly based on crosses between black passionfruit and yellow-fruited strains of black passionfruit (*P. edulis* f. *flavicarpa*).

Scions for commercial growers are sourced from field plantings of selected hybrids.

For home growers, nurseries may use young seedling black passionfruit as scion sources, in an attempt to delay infection with woodiness virus.

To grow seedling passionfruit, scoop out the seeds and clean them by fermentation for three days. Rinse, and sow immediately, or dry in the shade for 24 hours if you wish to store seed. Seed will remain viable in storage for about a year if kept dry and cool. Sow seed thinly, 5–10 mm deep, in a free draining medium. Keep damp but avoid watering too frequently.

Seedlings started in spring should be large enough to graft (3–4 mm diameter) at two to four months old, or earlier in subtropical areas. Grafting is possible any time during the warmer months.

Collect scion wood shortly before use. Cut pieces of green wood of a size to match the rootstock diameter, each with several buds and one leaf retained. If stored for more than a day or so, hold scions in an upright position to prevent curling of the tips as they continue to grow away from gravity. Store at about 12°C, either with the cut bases in water, or sealed in plastic bags.

Graft with a cleft graft, either on the top or on the side. For a top cleft, behead the seedling rootstock at a height where the diameter matches that of the scion, leaving at least several leaves below. The side cleft technique is illustrated. Note the leaf retained on the scion.

Figure 43. Seedling black passionfruit seedlings ready for use as scions.

Figure 44. The graft cut on a passionfruit rootstock.

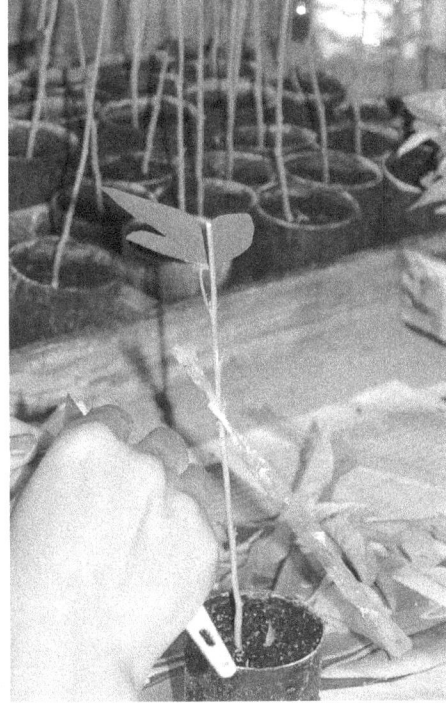

Figure 45. Cutting the scion.

Figure 46. Taping graft.

Figure 47. Aftercare of grafted passionfruit.

Bind the scion in place with a strip of paraffin wax film wrapped around the stem. Thin strips of budding tape are another alternative, but wrapping with tape requires some dexterity with these thin stems. Figure 46 shows a short piece of budding tape tied around the stem to hold the graft in place. The union is then sealed with a small piece of warmed, softened paraffin wax, squeezed around the stem.

After grafting, either cover the graft with a small plastic bag or hold the grafted plants in an area of very high humidity, such as a mist house.

In either case, shade the plants, but not too heavily (say 30%). The scions usually start growing three to four weeks later. At this stage, move the plants out of the protected conditions, or cut the top off the plastic bag.

Plant out in the field one to two months later.

Pistachio

In the USA, pistachio (*Pistacia vera*) is usually propagated by T-budding during the growing season. In Australia, the practice has been to chip bud in late spring. Both techniques have given variable success rates. Some Australian nurseries have found grafting to be more reliable.

Collect dormant scion wood for chip budding or grafting during late winter.

As pistachio trees bear their male and female flowers on separate trees, you will need to collect scion wood of both sexes. In an orchard, male trees are required in a ratio of about one male tree for every nine females.

Mature, nut bearing pistachio trees produce mainly flower buds and only a few of the vegetative buds that you need for budding and grafting. You will be able to recognise flower buds easily, as they are at least twice the size of vegetative buds (Figure 48). The best sources for suitable scion wood are either young trees before they begin to flower much, or older trees that have been heavily pruned to induce vegetative growth. If the

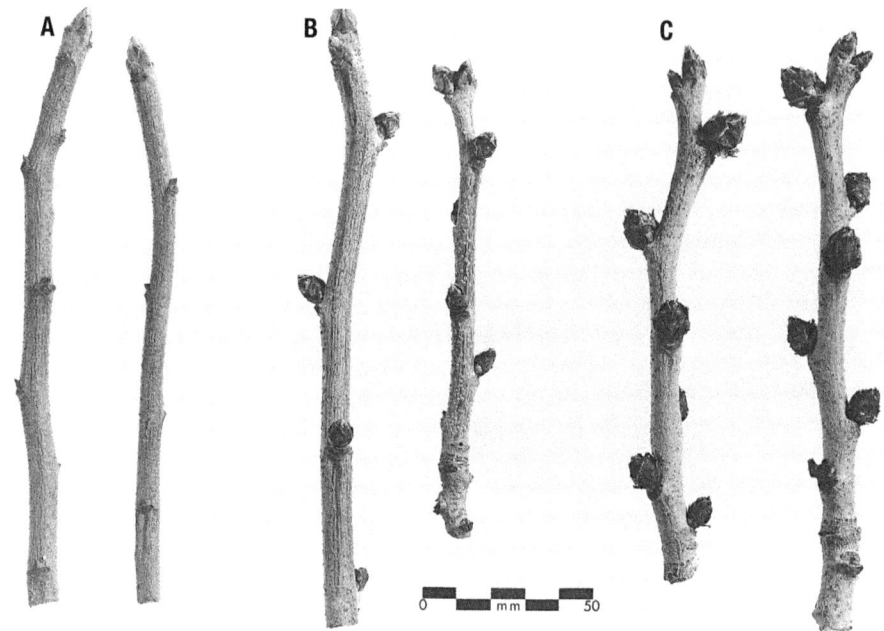

Figure 48. Pistachio shoots. A) Vegetative buds suitable for scion wood. B) Female flower buds. C) Male flower buds. Note that the terminal bud is vegetative in each instance.

regrowth shoots from these pruned trees are too thick for the available rootstocks, or if they begin to produce flower buds, prune them a second time in late spring. For propagation, select buds that are plump and well developed and not immature or floral (Figure 48).

Dormant scion wood may be stored under refrigeration at 5°C in sealed plastic bags for about six months.

Pistachio is not normally worked onto *Pistacia vera*, but onto seedling rootstocks of other related species. Current practice in Australia is to use either *P. terebinthus* or *P. integerrima* 'Pioneer Gold'. The latter has come into widespread use in the USA because of its tolerance of verticillium wilt, a common soil-borne fungal disease.

Chip budding with dormant wood in mid-spring has been moderately successful, but grafting with dormant wood at this time gives the best results. Field-planted rootstocks should be 15–20 mm in diameter, in active growth with some mature leaves. Pistachios can also be propagated onto rootstocks kept actively growing under glasshouse conditions.

Wrap buds to leave their tips exposed so that growth is not restricted. At the time of budding, prune off the upper third of the rootstock above the bud, but leave at least 10 leaves to ensure that the root system remains healthy.

In February and early March, field T- or chip budding can be done, using freshly cut green bud wood. For success with these techniques, selection of plump buds is essential, and the fresh bud wood should be used as soon as possible after it is collected.

After budding, semi-cincture the rootstock shoot with a single knife cut 2 cm above the inserted bud. The partial cincture overcomes the strong apical dominance of pistachio and encourages the scion bud to burst.

Shoots will develop from the buds after about three weeks. If exposed to wind, support the shoots as soon as possible by tying them to the rootstock. About two months after budding, cut the tape on the side of the rootstock opposite to the bud. Cut only the basal turns of tape, leaving the upper section of the tape intact to help restrict growth of the top of the rootstock. Remove this remaining part of the rootstock above the bud after nine to 12 months with a sloping cut about 15 degrees below the horizontal, almost flush with the new shoot.

Pome fruit (apples and pears)

Select bud wood from exposed vegetative shoots, with a diameter slightly smaller than that of the available rootstocks. Collect it either during the dormant season or in late spring or early autumn. Bud wood collected in the dormant season may be stored for use on actively growing rootstocks in spring, or may be used to graft rootstocks just before they start growing.

The wood may also be kept for budding in autumn, but at this time fresh bud wood is available. Select buds that are fully mature and plump.

For grafting dormant scion wood on dormant rootstocks, a whip, whip and tongue, cleft, or side graft is used. For budding in late spring or early autumn, a T-bud or chip bud is preferred. In late spring, either dormant bud wood or, if sufficiently mature, buds from new growth may be used.

Apple rootstocks are either seedling apple (for example, Delicious) or one of the newer dwarfing East Malling clonal apple stocks. Pears are worked either onto seedling pears, or for dwarfing, onto specially selected clonal quinces.

Roses

Rose scion varieties are mostly complex hybrids involving several species. These have been crossed and re-crossed over the last hundred years or so, resulting in our familiar many-petalled, often scented, repeat blooming roses, in almost every colour.

Most roses sold today are budded on to rootstock varieties chosen mainly for their vigour, which leads to strong scion growth, resulting in a saleable plant at one to two years of age. Rapid growth continues after planting out in the garden, quickly giving a good sized flowering shrub.

Tolerance to rootknot nematodes can be a consideration in lighter soils. Resistance to some soil-borne fungal diseases such as verticillium and crown gall may be important in some locations.

Another important reason for budding roses is the economy of propagating material, as theoretically you need only one scion bud to produce each saleable plant.

Other factors affecting choice of rootstock for roses include some horticultural characteristics of the rootstock itself, such as ease of striking cuttings, lower thorniness, bud take and less of a tendency to sucker.

Despite all these reasons for budding roses, groundcover roses and miniature roses usually grow well on their own roots. Other vigorous varieties may also grow well on their own roots, where nematodes are not a problem.

In Australia, you will find different rootstocks used according to State. Victorian nurseries prefer *Rosa multiflora* 'Kiwi', a strong, almost thornless variety that grows very well in the absence of nematodes. 'Kiwi' is particularly suitable for the production of the long shoots needed to make 'standard' roses.

NSW and South Australian nurseries mostly use 'Dr Huey', a modern hybrid climbing rose, because of its heat tolerance and superior performance on alkaline calcareous soils. 'Dr Huey' has a few thorns, and is a little more likely to send up rootstock suckers, especially in the first few years. Both *R. multiflora* and 'Dr Huey' are easy to strike as cuttings, are easy to bud, and neither suckers badly, provided the cuttings are disbudded.

Western Australians rely on *R. fortuniana* for its better nematode tolerance and superior vigour in their sandy soils. *R fortuniana* is a little more difficult to strike as cuttings, but buds satisfactorily as long as the bark is slipping.

Rosa indica is sometimes used as a rootstock for roses grown for cut flower production.

Although virus tested material is available for some rootstock selections, there may be little benefit in using it, since there are very few virus-tested scions in Australia.

Production of a budded rose for sale

Take cuttings of rootstock varieties from specially grown mother plants in autumn. Cut to appropriate lengths, as follows: about 15–20 cm for bush roses; about 70 cm for 'patio' standard roses; and about 1 m for regular standard roses.

De-thorn if necessary and disbud the cuttings, leaving only two buds at the top of the cutting.

Cuttings may be callused by burying in a shallow trench for about three weeks.

Plant callused cuttings in late autumn to early winter. Cuttings will take root and commence new shoot growth by early spring.

Collect bud wood from mother plantings in mid to late winter, carefully identifying each variety with permanent labels. At this time of year, any part of the current season's shoots make suitable bud wood. De-leaf and de-thorn the shoots and cut into convenient length pieces, say 15–20 cm. Store bud wood in cool, humid conditions, for example in plastic bags in a cool room.

Start budding in late spring to early summer, when the new shoots on the rootstock cuttings have hardened off. Bud by the T-bud method, and bind and cover the bud with a rubber strip or patch. After about three weeks, cut off the top of the rootstock just above the bud, so that the bud is forced into growth. In sunlight, the rubber patch deteriorates sufficiently so that either the bursting bud easily grows through the patch, or it falls off. Under cooler, cloudy or shaded conditions, you may have to remove the patch to let the bud grow.

Grow the plants on for the rest of the season, and dig from the field in late autumn for sale as one year old roses. For two year old roses, bud in late summer with fresh bud wood, and dig late in the second autumn.

In commercial field production, two man teams often perform the budding. The 'budder' carries a supply of bud sticks and a knife. He moves down the row, working bent over, making the T incisions in the rootstock stems, cutting the bud from the bud stick and inserting it in the T (Figures 49–53). The 'patcher' follows behind, covering each bud with a pre-stapled thin rubber patch (Figure 54).

Sapodilla

This species exudes a sticky white latex whenever the bark is cut or damaged, so special techniques are necessary for its successful propagation.

Because sapodilla is evergreen, you can bud or graft at any time, provided the rootstock is actively growing, and suitable scion wood is available. Use a terminal whip and tongue graft, a side graft or a patch budding technique.

Figure 49. Because of the biomechanics of the situation, the budder makes the first cut of the 'T' diagonally, rather than across the stem.

Figure 50. Second cut, the stem of the 'T' and the bark lifted simultaneously.

Figure 51. Cutting the bud.

Figure 52. Bud inserted.

Figure 53. Bud tail cut off.

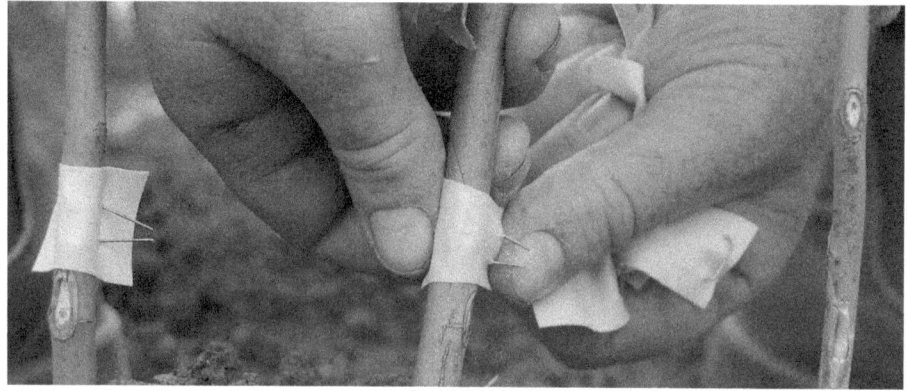

Figure 54. Rubber patch in place.

For grafting, select a mature main shoot as the scion source, and remove the leaf blades from the last 5 cm. After approximately three weeks, the petioles of the disbladed leaves will have fallen off and the scars healed. Five or ten minutes before collecting the scion wood, slash the bark of the scion shoot at several places below the prepared part with long, diagonal cuts. At the same time, sever the rootstock just above the intended position of grafting and similarly slash its bark 5–10 cm lower down. Later, when you make the graft, much less latex oozes from the surfaces of the grafting cuts.

For a patch bud, release the latex pressure in scion and rootstock as described for grafting. Cut the patch around the bud and pull the patch sideways from the stem without bending it. Wrap the patch bud, then semi-cincture the stem above the bud to encourage bud burst.

Prune back the top of the rootstock by about one-third at the time of budding, and further reduce it when the bud begins to grow. The removal of leaves on the side of the rootstock carrying the bud will sometimes assist the development of the scion.

Stone fruits

Budding is the preferred technique for propagation of stone fruits. Bud wood may be collected while the tree is dormant, or during the growing season. Select buds that are pointed rather than rounded, since the latter are usually flower buds. If there are multiple buds, the central one should be pointed, which indicates that it is vegetative. The outer two floral buds may then be rubbed out.

Bud wood collected in the dormant season may be stored for several months, but when collected during the growing season, it should be used as soon as possible.

The rootstock should be actively growing so that the bark lifts readily. A T-bud may be used in summer or autumn, and the most suitable time ranges between December and March for different species as shown in Table 1.

Table 1 Preferred budding times for stone fruits

Species	Best budding time
Almond	Late December, January
Apricot	January
Plum	February
Nectarine	February
Peach	February, March

There is also a suitable time to bud in spring. At this time, bud wood from new growth on the source trees is generally too immature for satisfactory budding, so buds from stored dormant bud wood are used.

At the time of budding, it is common practice to remove the terminal third of the rootstock. The remainder of the rootstock above the bud may be used as a support for the developing scion, but any suckers originating from the stock must be removed to prevent the scion from being shaded out.

If the rootstock is wounded too deeply on some species, a gum is exuded which prevents normal growth of the bud. Therefore, take care that the T is cut only just through the bark.

Almond rootstocks are usually almond seedlings or sometimes peach seedlings if nematodes are a problem.

Peaches and nectarines are generally worked onto peach seedlings, or plum for heavy, waterlogged soils.

Myrobalan plum (cherry plum) seedlings make a satisfactory rootstock for both European and Japanese plums.

Tomatoes

Tomatoes (*Lycopersicum esculentum*, syn. *Solanum lycopersicum*) originated in the Americas, and did not come into general use in Europe until some time between the mid 15th and 16th centuries. Tomatoes are now one of the most popular commercial and home garden vegetables in temperate climates. Unfortunately, they are subject to a variety of soil-borne pests and diseases. There are several ways to solve these problems. One is to breed resistance into new tomato varieties, using resistant wild types. Most of the newer varieties have multiple resistances to strains of the soil-borne fungus genera *Verticillium* and *Fusarium*. Nematode tolerance is also usual in modern tomatoes varieties. Some varieties may also include resistance to virus diseases, such as tobacco mosaic and tomato spotted wilt. However, breeding a new tomato variety that produces good quality fruit plus all these resistances takes considerable time and effort, and sometimes resistance becomes useless as new strains of pathogens develop.

A more flexible approach is to graft seedling-grown, good tomato varieties onto seedling rootstocks possessing these same traits of resistance to soil-borne fungi and nematodes. These rootstocks are usually purpose-bred, hybrid tomato varieties that do not have to produce commercially acceptable fruit. Such rootstocks (for example 'Beaufort') now are used even for large-scale production of grafted plants for commercial tomato growing. Sometimes the rootstock may be a modern tomato variety that has proven resistance, or a primitive tomato type that is small fruited but has reasonable resistance. Occasionally other wild *Solanum* species may be used.

Starting in late winter, first sow seed of the scion variety or varieties in seed raising mix in punnets or plug trays. Germinate seed in the dark, in a heated house, or with bottom heat, so that the temperature at seed level is about 25°C. Germination will occur after a week or less. Three to five days after sowing the scion seed, sow seed of the rootstock variety under similar conditions. The delay in sowing the scion seed will result in the scion and rootstock seedlings being the correct relative sizes for successful grafting at about two to three weeks old, 15 cm or so in height.

To top graft, behead the rootstock seedling below the seed leaves, using a sterile razor blade to make a cut at 45 degrees across the stem (Figure 55). Discard the top. Apply a grafting clip or tube around the top of the cut rootstock stem (Figure 56). Similarly, behead the scion seedling at 45

Figure 55. Making the grafting cut on a cutting board. The plants are still in a punnet.

degrees, also below the cotyledons, at a point where it matches the rootstock stem in diameter. Insert the cut lower end of the scion into the clip or tube so that it matches with the cut upper end of the rootstock. Hold the freshly grafted plants in a shaded humid environment for about a week, until the graft heals. Regular fogging or misting helps to maintain humidity. Gradually harden off the plants over several weeks, and move to an unshaded greenhouse. Pot on into 10 cm pots for sale two or more weeks later.

To approach graft, sow the scion and rootstock seeds in separate plug trays, at the same time. Germinate seeds in the dark, with bottom heat. Pot on the seedlings into punnets at three weeks, and return them to bottom heat. About a week later, transfer the punnets into an unheated greenhouse so that the plants can harden off before grafting.

At grafting time, remove the rootstock plant from its punnet and behead it square across the stem, just above the first pair of true leaves.

Make a downward sloping cut into the rootstock stem, starting just below the remaining leaves and running 15–20 mm down the stem and about half way through it (Figure 57).

Figure 56. Completed top graft held with clip.

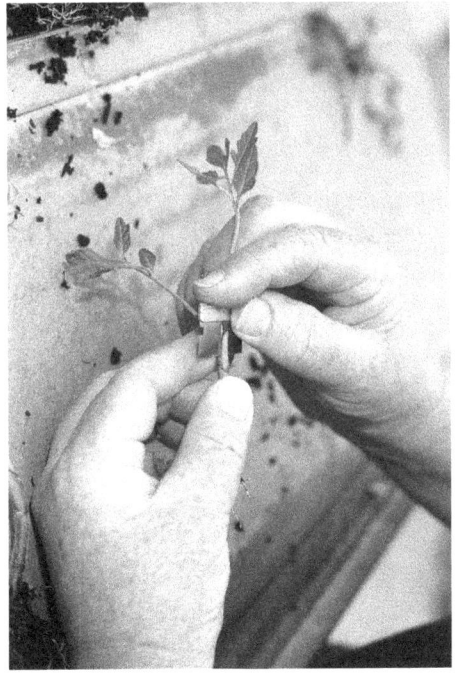

Figure 57. Approach graft of tomato. The first grafting cut on the beheaded rootstock.

Figure 58. The grafting cut on the scion plant.

Figure 59. Scraping the stem to expose more cambium.

Figure 60. Completed graft, pegged together.

Remove the scion seedling from the punnet. Starting about 20 mm below the first set of true leaves, make an upwards sloping cut in the scion stem to match the one made in the rootstock (Figure 58). Then, starting just below the leaves, scrape off the skin along the stem down to the beginning of the grafting cut. This will expose more of the scion's cambial regions for union with the cut surface of the rootstock (Figure 59). Similarly, scrape the skin off the rootstock stem, starting from the grafting cut and going down the stem for a centimetre or so.

Fit the scion and rootstock grafting cuts together, and hold in place with a grafting peg (Figure 60). No other binding or sealing is needed.

Re-pot the two plants into the one pot, stake and tie the scion, and hold the plants in a shaded, humid location over bottom heat for two weeks until the graft area has callused over. About 8–10 days after grafting, prune off the remaining top of the rootstock just above the graft. It is not necessary to sever the scion's own root system.

Two weeks after grafting, move the plants from bottom heat into an open greenhouse.

Regularly check the grafted plants, and remove any rootstock suckers as they arise.

Plants are ready for sale three weeks later.

Eggplant (*Solanum melongena*) can be propagated in a very similar fashion, using the same rootstock.

Walnut

Walnut (*Juglans regia*) is one of the most difficult trees to bud or graft, perhaps due to its rather slow rate of callus formation except under consistently warm weather conditions.

Collect scion wood during the dormant season, just before bud swell. It may be stored under refrigeration for up to six months for use after the rootstocks have produced their first fully developed leaves. For autumn budding, cut and use fresh scion wood from current season's growth.

On mature fruiting trees, most of the buds are flower buds, which are unsuitable for propagation. For this reason, you should severely prune bud wood source trees to keep them in a vegetative state. Alternatively, you can prune one or more branches of a mature tree. Two-year-old scion wood gives results that are more reliable.

Some suitable rootstocks for walnut are seedlings of the related Californian black walnut (*Juglans hindsii*) and hybrids between this species and walnut. For the best chance of success in budding or grafting, keep the rootstocks actively growing, well watered and well fertilised. Try to graft when daily minimum temperatures are over 15°C, and the daily maximum is reaching about 30°C.

Begin grafting in spring, after the rootstock has produced its first fully developed leaves. Use a whip and tongue or side graft, depending on the relative diameters of the rootstock and scion.

Patch bud (see Figure 15) either in spring with stored dormant wood, or in autumn with fresh wood. Patch budding is preferred because walnut has a very thick bark and is difficult to T-bud. Before patch budding with cool-stored dormant wood, stand it with its base in water at 20°C for a few days, until the bark is slipping.

You can improve the 'take' of walnut and other hard-to-graft species by supplying heat directly to the graft union area only. Splice graft cool-stored dormant rootstocks which have been grown in tubes or bare-rooted. Lay the graft unions across a narrow box heated to 27°C and cover. Keep the top of the scion cool and keep the roots cool and moist. After three to four weeks, when the buds start to burst, remove the grafted plants from the heat treatment box, and pot on or plant out.

Propagation of pecan (*Carya illinoiensis*) is very similar to that of walnut, although 'take' is usually better.

Ziziphus

Two species of *Ziziphus* are cultivated for their fruit. The Chinese jujube (*Z. jujuba*) is a deciduous shrub widely grown in China for its sweet fruit, which can be eaten either fresh or dried. The Indian jujube (*Z. mauritiana*) is a very productive, thorny, evergreen large shrub, popular in the warmer parts of India. *Ziziphus* species are declared noxious weeds in northern Australia.

Grafting is advised rather than budding for the Chinese jujube. It has very thin bark that rarely separates cleanly from the wood without splitting down the stem.

Graft Chinese jujube in early spring, using dormant wood, or in late summer and autumn using prepared green graft wood. Collect graft wood only from vigorous, vertical shoots. The thin, lateral fruiting shoots fall in autumn and will die at that time if used as summer scions. If you use the old, larger laterals as scions, they tend to produce only these small deciduous shoots and can be slow to make extension growth. Store

Methods for selected species | 93

Figure 61. Chinese jujube 'Li', in fruit.

dormant graft wood in a sealed plastic bag under refrigeration and graft it in early spring, just before bud burst of the rootstock. Green graft wood can be prepared as previously described and used fresh.

Glossary

abcise	To separate from naturally and fall off, like leaves in autumn.
axil	The place between a shoot and the upper side of a leaf stalk (that is, the side of the leaf stalk away from the roots of the plant).
budding	Grafting with only a small piece of shoot carrying a single bud.
budding tape	Plastic tape for wrapping buds and grafts to hold and seal them. Usually in rolls 12 mm wide.
callus	Scar tissue formed by plants when healing wounds, including grafts.
cambium	The part of plants capable of forming new tissue, including scar tissue in response to wounding.
cincture	To remove a narrow ring of bark from a shoot or branch.
clone	A group of identical plants, or one of this group.
deciduous plant	A plant which stops growing and loses its leaves in autumn, then produces new leaves and shoots in spring.
disblading	Cutting off leaf blades, but leaving the leaf stalks attached to the shoot.
dormant	Not growing and leafless, as are deciduous plants in winter.
grafting	Taking from a selected plant a length of shoot with some buds, then attaching it to another plant so that they grow together.

latex	A sticky, milky liquid which oozes from some plants (e.g. figs, rubber plants) when they are cut.
mature	Fully grown and no longer soft and tender.
multi-graft	A single rootstock plant grafted with more than one variety of scion.
nematode	A group of microscopic worm-like organisms, some of which live in the soil and attack plant roots.
pare	To cut off a very thin slice, in order to tidy up a rough secateur cut or to improve an unsatisfactory grafting cut.
petiole	The stalk by which the leaf blade is attached to the shoot.
rootstock	The plant that provides the roots and a short part of the trunk of a budded or grafted plant; that part of the budded or grafted plant.
scion (1)	A piece of shoot (including at least one vegetative bud) taken from a selected plant to propagate a new plant by budding or grafting.
scion (2)	The part of a grafted or budded plant above the graft union (trunk, branches, etc.).
sucker	A strong shoot coming from low down on a plant, possibly from the rootstock on a grafted plant.
vegetative propagation	Growing a new plant from part of a shoot or root of a selected plant. The new plant is identical to the parent plant.

Further reading

Allen A (1986) *Growing Nuts in Australia*. Night Owl, Shepparton.

Baxter P (1998) *Growing Fruit in Australia: For Profit or Pleasure*. Pan Macmillan, Melbourne.

Bose TK (2005) *Propagation of Tropical and Subtropical Horticultural Crops*. Naya Udyog, Kalkata, India.

Garner RJ (1988) *The Grafter's Handbook* (5th edn). Cassel, London.

Glowinski L (1999) *The Complete Book of Fruit Growing in Australia*. Hachette Livre, Sydney.

Hartmann HJ, Kester DE and Davies FT (1997) *Plant Propagation: Principles and Practices* (7th edn). Prentice Hall, Upper Saddle River, New Jersey.

Macdonald B (1986) *Practical Woody Plant Propagation for Nursery Growers*. Timber Press, Portland, Oregon.

Internet resources

The Aggie Horticulture pages of the Texas A & M University.
http://aggie-horticulture.tamu.edu/propagation/propagation.html

The American Rose Society pages: 'Hints for Successful Grafting and Budding'.
http://www.ars.org/About_Roses/propagating-hints.html

The Association of Societies for Growing Australian Native Plants: 'The Propagation of *Banksia*'.
http://asgap.org.au/APOL24/dec01-1.html

The Australian National Botanic Gardens: 'Grafting Australian Native Plants'.
http://www.anbg.gov.au/hort.research/grafting.html

University of Georgia extension pages
http://pubs.caes.uga.edu/caespubs/pubcd/B818.htm

University of Missouri extension pages
http://extension.missouri.edu/xplor/agguides/hort/g06971.htm

Index

Abies   64
Acer   65
almond   86
ash   65
Annona species   47
apple   79
apple, crab   67
apricot   86
atemoya   47
avocado   49

Banksia   41
Betula   66
birch   66
bullock's heart   47

cacti   52
Carya   92
cashew   56
Casimiroa   49
Cedrus   62
Cereus   53
cherimoya   47
Corymbia   40
cherry, ornamental   66
cherry plum   86
Chinese jujube   92
citrange   47, 59
citrus   14, 19, 23, 59
Citrus australasica   47
Citrus glauca   47
Cleopatra mandarin   59
conifers   61
Cupressocyparis   62
Cupressus   62
custard apple   47

Darwinia   46

desert lime   47
Diospyros   49
Eucalyptus   39

elm   68
Eremophila   46
eucalypts   39
Eucalyptus   39
Euphoria longan   49

finger lime   47
Fraxinus   65

Gleditsia   67
grape   68
grapefruit   *see* citrus
Ginkgo   64
Grevillea   44
Gymnocalycium   52

Hakea   44
honey locust   67
Hylocereus   53

jackfruit   20
Juglans   91
jujube   92
Juniperus   61

lemon   *see* citrus
lime, West Indian   61
longan   49
Lophophora   53
Lycopersicum   86

macadamia   70
Malus, ornamental species   67
Mammillaria   53

mandarin   59
mango   71
maple   65
Mazzard   66
mint bush   46
Moon cactus   52
Mop Top (*Robinia*)   67
Myoporum   46
Myrtillocactus   53
myrtle cactus   53

nectarine   86

Opuntia   54
orange   *see* citrus

passionfruit   73
Passiflora   73
peach   85
pear, edible fruiting   79
pear, ornamental   67
pecan   92
Pereskiopsis   53
persimmon   49
Picea   62
Pinus   62
pistachio   77
Pistacia   77
plum   85
pome fruit   79
Poncirus   59
prickly pear   54
Prostanthera   46
Prunus, edible fruiting   *see* stone fruits

Prunus, ornamental   66
Pyrus, ornamental   67
Pyrus, edible fruiting   *see* pome fruit

quandong   46
quince   80

Robinia   67
Rosa   80
rose   14, 80

Santalum   46
sapodilla   82
Solanum   86
soursop   47
stone fruits   85
sweetsop   47
sweet orange   3, 59

Taxodium   64
Thuja   63
tomato   86
torch cactus   54
Trichocereus   54

Ulmus   68

Vitis   68

walnut   91
Westringia   46

Ziziphus   92

www.ingramcontent.com/pod-product-compliance
Ingram Content Group UK Ltd.
Pitfield, Milton Keynes, MK11 3LW, UK
UKHW062045180426
11947UKWH00030B/2049